Applications of Mass Spectrometry to Organic Chemistry

Applications of Mass Spectrometry to Organic Chemistry

R. I. REED
Chemistry Department
The University
Glasgow
Scotland

1966
ACADEMIC PRESS London and New York

ACADEMIC PRESS INC. (LONDON) LTD
Berkeley Square House
Berkeley Square
London, W.1

U.S. Edition published by

ACADEMIC PRESS INC.
111 Fifth Avenue
New York, New York 10003

Library of Congress Catalog Card Number: 65–27319

Printed in Great Britain by
Spottiswoode Ballantyne and Company Limited
London and Colchester

Preface

The development of mass spectrometry in its application to organic chemistry has been somewhat hesitant. At first, by reason of the numerous re-arrangement processes observed, it was doubted if the method would be of value in the analysis of organic structures, always excepting the contribution made in the Petroleum Industry. Then, thanks largely to the pioneering efforts of a few, interest in this application was awakened. Now the development of new instruments and techniques has so enlarged the horizons that, allowing for the ingenuity of the workers, the subject seems without bounds.

The present approach to this study seems to follow two main lines. Attempts are being made to deduce the structure of the organic molecule from the observed mass spectrum, and, conversely, a *rationale* is provided for the fragmentation pattern of a known structure. Several textbooks are available for these purposes. The present book stands somewhere in between. The first part deals with the information necessary to provide an understanding of a mass spectrum and enable valid deductions to be made therefrom; the second part outlines one method in which the mass spectrometer may be used as an adjunct to chemical manipulation. In the relatively simple structures examined, and in order not to obscure the mass-spectrometric considerations, the fewest possible chemical operations are employed. However, for the more experienced worker, more sophisticated chemical reactions may be considered.

In the writing of this book I have benefited greatly from the comments of my friends and my research students. I am also indebted to Dr. N. McCorkindale of The University, Glasgow, for examining the first part of the text and for the subsequent benefit of much discussion. His comments have done much to clarify many points in the argument. I am also most grateful to Professor G. R. Hill of the University of Utah, in whose Department much of the second part was developed.

I am indebted to Academic Press for the kind and courteous way they have helped me prepare the work, and for the fine production of the volume itself.

Glasgow
August 1965

R. I. Reed

Contents

CHAPTER 7

The Analysis of Spectra

CHAPTER 8

Logical Argument

CHAPTER 1

Instrumentation

A. Instrument Design

Many kinds of mass spectrometer exist and these are adequately reviewed in the existing literature (Barnard, 1953, p. 152; Beynon, 1960, p. 5; Elliott, 1963; Zahn, 1963). The majority of them form ions by electron bombardment, and sort the ions so obtained into beams of positively charged particles which are homogeneous in the mass/charge (m/e) ratio. Since only a few of the ions are formed bearing two or more positive charges, the mass/charge ratio provides a simple means of determining the mass of the ion under examination. This resolution is achieved by electrostatic repulsion followed by magnetic deflection. The formula appropriate to the combined effects is

$$\frac{m}{e} = \frac{r^2 H^2}{2V}$$

where m is the mass of the ion of charge e, r is the radius of gyration of the ion, H is the field strength of the magnet, and V is the electrostatic potential. From this equation it follows that ion beams differing in m/e values can be made to pass through a fixed slit beyond the magnetic field by varying either V or H. In Dempster type instruments, in which the ions describe a 180° arc, it is customary to maintain H constant and vary V. In the sector type instruments V is held constant and H varied.

Instruments may be further divided into single and double focusing; single-focusing designs are discussed in the literature listed above. The ions after electrostatic repulsion enter the magnetic field and are brought to a focus at a slit, the exit slit, immediately in front of the collector (Fig. 1). Such instruments possess two slits, one beam-defining in the ion source, and the exit slit. Since a magnetic field has some focusing properties; the ion beam, when correctly focused, produces an image of the beam-defining slit at the exit slit.

In order to be of value, the adjacent beams of mass m/e and $(m+1)/e$ must be separated one from the other, i.e. resolved. Clearly, it is not necessary, for the average organic application that the two beams be distinct over their entire length on a record; some overlap may occur.

The extent of this is used to define resolving power. Different authorities have different criteria, but it is convenient to employ here the definition commonly used by the instrument manufacturers in Great Britain. Two adjacent ions of equal abundance are said to be resolved if the overlap between them does not amount to more than a 5% contribution from each (Fig. 2). The resolving power of any instrument will depend upon many factors: e.g. the slit width, and the field strength of the magnet. There is also the trajectory of the ions formed. These ions may follow a divergent path on leaving the beam-defining slit and, moreover, will contain some ions which, for various reasons, do not possess the kinetic

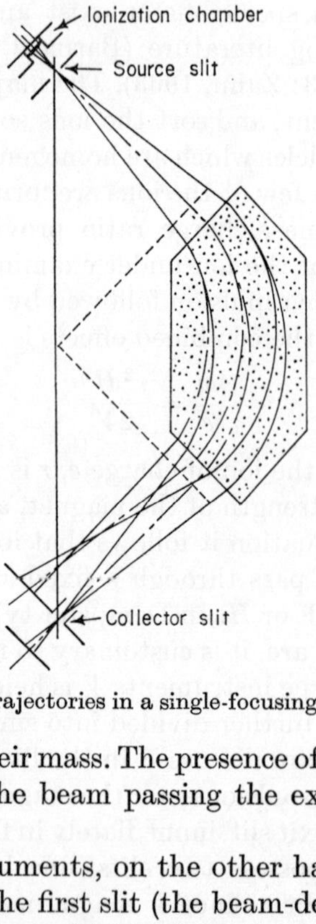

FIG. 1. Ion trajectories in a single-focusing instrument.

energy appropriate to their mass. The presence of such ions broadens and modifies the shape of the beam passing the exit slit, and thus limits resolution.

Double-focusing instruments, on the other hand, possess three slits. The ions, after passing the first slit (the beam-defining slit), are electrostatically deflected and brought to a focus upon a second (monitor) slit. This focused beam becomes the image for the magnetic deflection (Fig. 3). The electrostatic focusing excludes from the magnetic analyser all the ions which do not have the appropriate kinetic energy. The resolved ion beam is, therefore, much better defined, and a higher resolution is obtainable. One commercially available instrument using a

12-in. radius of gyration has been reported to give a resolution of five parts in 10^5 or better.

Both types of instrument are of considerable importance in the

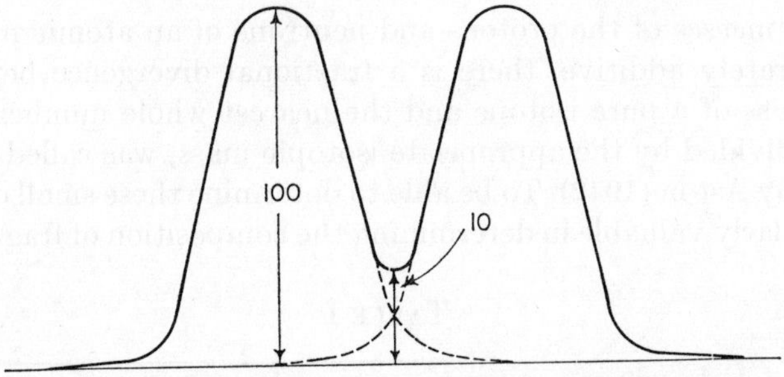

FIG. 2. Two fully resolved adjacent peaks on the basis of the "10% valley" definition.

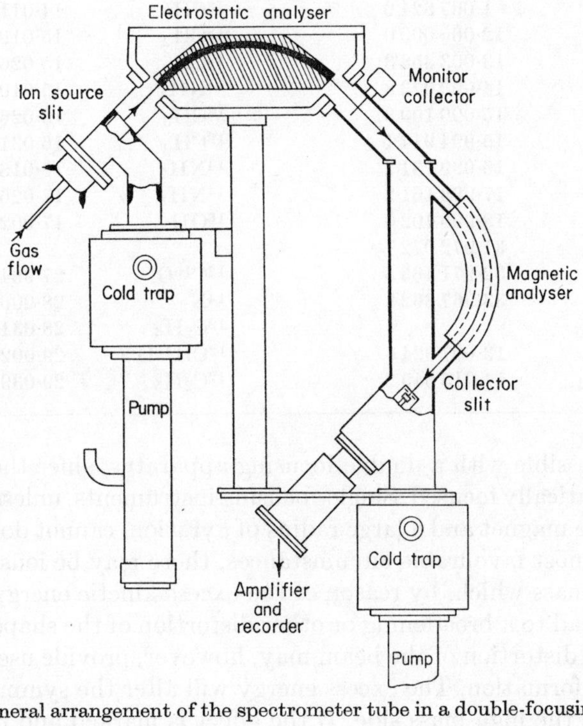

FIG. 3. General arrangement of the spectrometer tube in a double-focusing instrument of Nier-type geometry.

applications of mass spectrometric analysis to organic compounds. The high resolving power of the double-focusing instrument is invaluable, because it enables the precise mass of ions to be determined. The exact atomic weights of the common isotopes (referred to carbon-12, 12·000000) are not integers and, provided that the mass measurements are precise enough, it is possible to distinguish species of the same nominal mass (Table I).

As the masses of the protons and neutrons of an atomic nucleus are not accurately additive, there is a fractional divergence between the actual mass of a pure isotope and the nearest whole number. The difference, divided by the appropriate isotopic mass, was called a packing fraction by Aston (1942). To be able to determine these small differences is particularly valuable in determining the composition of fragment ions,

Table I

Species	Mass	Species	Mass
^{1}H	1·007 824 6	^{13}CH	14·011 182 9
^{12}C	12·000 000 0	$^{13}CH_2$	15·019 007 5
^{13}C	13·003 358 3	$^{12}CH_3$	15·026 473 8
^{14}N	14·003 073 8	^{14}NH	15·010 898 4
^{15}N	15·000 109 8	$^{13}CH_3$	16·026 832 1
^{16}O	15·994 914 1	$^{12}CH_4$	16·031 298 4
^{17}O	16·999 131 2	$^{14}NH_2$	16·018 723 0
^{18}O	17·999 161 2	$^{14}NH_3$	17·026 547 6
^{19}F	18·998 402 0	^{16}OH	17·002 738 7
^{32}S	31·972 072 7		
^{33}S	32·971 463 5	$^{12}C^{16}O$	27·994 914 1
^{34}S	33·967 862 8	$^{14}N_2$	28·006 147 6
		$^{12}C_2H_4$	28·031 298 4
^{12}CH	13·007 824 6	$^{12}CH^{16}O$	29·002 738 7
$^{12}CH_2$	14·015 649 2	$^{12}C_2H_5$	29·039 123 0

and it is possible with a double-focusing apparatus, since the ion beam is electrostatically focused. Single-focusing instruments, unless they have a very large magnet and a large radius of gyration, cannot do this. Even under the most favourable circumstances, there may be ions present at any given mass which, by reason of the excess kinetic energy possessed by them, lead to a broadening or other distortion of the shape of the ion beam. This distortion of the beam may, however, provide useful supplementary information. The excess energy will alter the symmetry of the beam upon the high-mass side. If the effect is marked and if the beam can be compared with one which does not possess this extra translational energy, the effect may be used to detect, and possibly estimate, such energy. As an aid to analysis the method may prove rewarding. So far it has been unreasonably neglected.

B. Ion Detection

The customary method for the detection of ions, particularly in single-focusing instruments, is by valve amplification. The positive ion current, the input signal, is fed onto the grid of an electrometer valve. The output is further amplified by a d.c. amplifier, and displayed upon a meter, a pen-recorder or a galvanometer recorder. The last device is now the most commonly used; it comprises a set of mirror galvanometers which record the signals upon a suitably sensitized paper, allowing ions of very different signal strengths to be recorded on the same trace. Recently, multi-channel pen-recorders have also become commercially available, and these provide a suitable substitute for the galvanometers. The lower limit of detection of a signal depends upon the design of the valve detector and its associated circuits; a reasonable lower limit would be about 1×10^{-15} A. Vibrating-reed electrometers may be used as a substitute, and these are somewhat more sensitive.

When double-focusing instruments are operating under conditions leading to high resolving power, the slit widths employed are very small (some ten thousandths of an inch for the beam-defining slit), and the strength of the ion beam is consequently much smaller. The amplification of so weak a signal is necessarily greater, and the level of detection must be much lower. For these reasons, it is customary to use electron multipliers as the first stage in detection and amplification.

The very small ion currents which may now be measured introduce one other consideration. The low number of ions produced over short intervals of time may vary (the so-called statistical factor), which results in an irregular outline to the top of an ion beam when recorded by a galvanometer or displayed on the screen of an oscilloscope. Variations in the gas pressure of the sample examined, in the quantity of electrons emitted by the filament, changes in the potentials in the ion source, and many other factors may also lead to such variations.

An alternative method of amplification for instruments of low resolving power which has been but little used is to employ a.c. amplification (Beynon, Clough and Williams, 1958). A pulsed ion beam is generated at a given frequency and it is detected and amplified by an a.c. system tuned to the same frequency, when a derived, rather than the conventional, spectrum is obtained. That is the ions now recorded (Fig. 5) are a measure of the slopes of the sides of the ion beam received by the collector. When a phase-insensitive detector is employed, each ion appears twice over by comparison with the conventional spectrum, and the mass value of the ion occurs at the zero point between the two lobes of each peak. Any asymmetry existing in the ion beam now becomes apparent in the derived spectrum. Thus, it is much easier to detect ions possessed of excess translational energy, and the method may provide a simple way of recognizing ion beams so possessed, although, of course, there are other

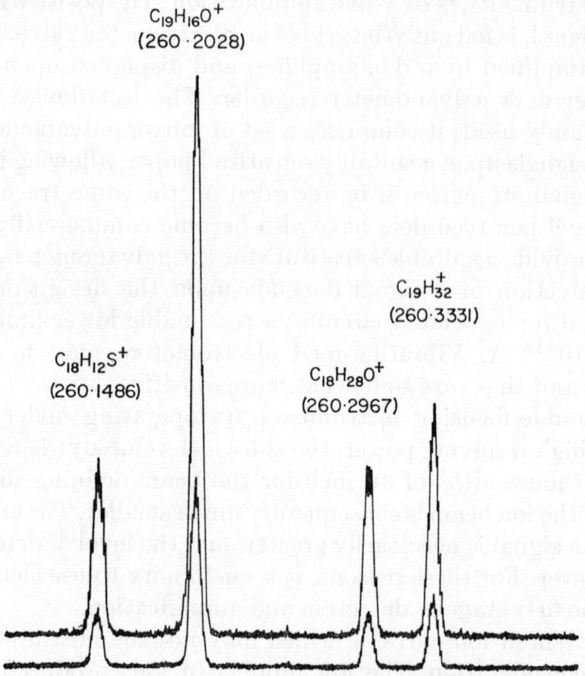

FIG. 4. A typical high resolution scan showing the multiplet at mass 260.

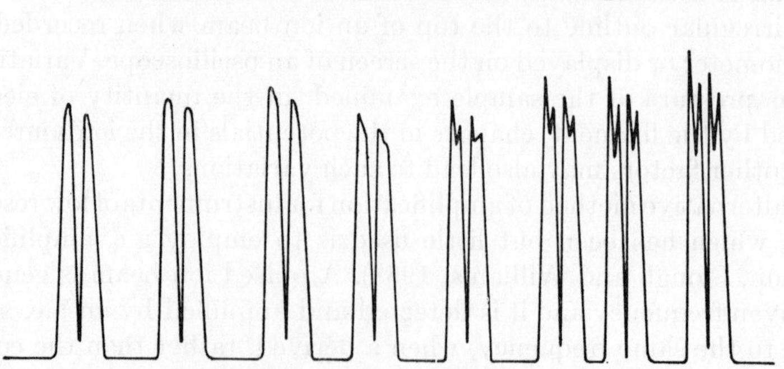

FIG. 5. First derivative spectrum. The four peaks to the left are singlets, the four to the right doublets.

conditions, arising from the unequal potentials in the ion source, which may produce the same effect. These are best eliminated by comparison with other, symmetrical ions, which are also obtained. Ions may also be produced by other means. One method is that of field emission spectrometry, which has been developed by Beckey (1963), Gomer and Inghram (1954, 1955) and others. The principle of operation is as follows. A high potential is applied to a very fine metal point so that a voltage of several volts per Ångstrom is developed. The tip is surrounded by vapour of the material under examination. Under these conditions, molecular ions of the compound are formed by electron extraction. As the bulk of the ions obtained are parent molecular ions, this leads directly to the molecular weight of the material. Because a knowledge of this is of great importance in organic chemistry and may be difficult to obtain by other methods, field emission methods of ion production will become of increasing importance. The argument necessarily supposes that the volatilization of the substance under examination does not decompose it entirely. Many other types of instrument have been made, some of which dispense with the magnetic analyser and depend only on electrostatic repulsion to sort the ions. So far none of these instruments has a resolution comparable with those already discussed, nor have they come into general use in the analysis of organic compounds. Accordingly, they are not discussed here and for details reference should be made to the original communications (Brubaker and Perkins, 1956; Goudsmit, 1948; Hipple, Sommer and Thomas, 1949, 1950; Sommer, Thomas and Hipple, 1951; Bennett, 1949, 1950a, b, 1953; Falk and Schwering, 1957a, b; Stephens, 1946; Wiley and McLaren, 1955; and Wiley, 1956).

C. Sample Admission

The sample to be analysed is usually admitted to the ion source as a vapour. In order to obtain reproducible spectra, therefore, the vapour pressure of the substance must be sensibly constant for the period of measurement. There is the further requirement that the pressure be kept low in order to minimize the likelihood of collisions between the ions and neutral molecules. Such collisions will destroy the simple ion optics upon which the analytical method depends, and may also lead to bimolecular reaction. These reactions will be discussed in a later chapter (see Chapter 2, C, p. 27). Because of the occurrence of ion–molecule reactions, most instruments are operated with a source pressure of 10^{-6}–10^{-7} torr (10^{-7}–10^{-8} cm Hg), under which operating conditions simple, reproducible spectra are obtained. The two pressure requirements are met by storing the vapour in a large reservoir of 1–2 litres capacity and leaking it into the ionization chamber through a fine sintered disc (Fig. 6). Relatively few classes of organic compound are gaseous at room temperature and, although many liquids have sufficient vapour pressure to be treated in this way, the majority of substances cannot be so handled. The introduction of a heated reservoir (O'Neal Jr. and Wier, 1951) greatly extended the range of compounds that could be examined. The reservoir, which may be of glass or metal lined with enamel, can usually be heated to 300° C, occasionally to 350° C. The sintered disc and the inlet line are also heated to the same temperature, as well as the ion source, in order to prevent condensation. The draw-back to this method is that some classes of compounds may suffer modification under the experimental conditions. Many alcohols, particularly of high

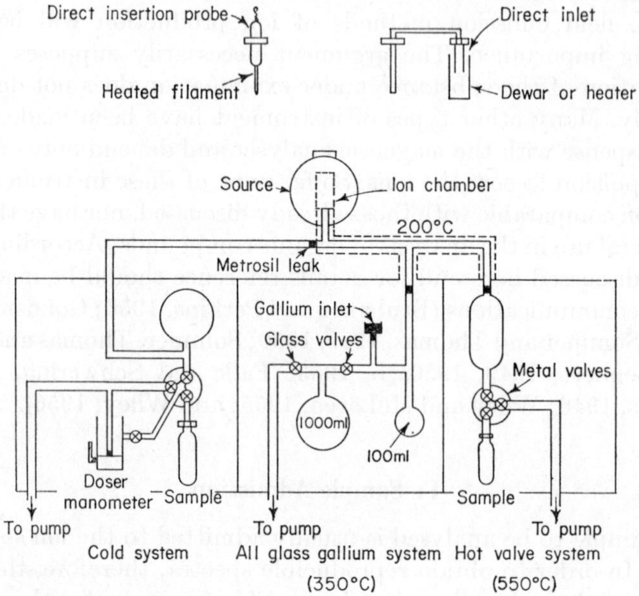

FIG. 6. Schematic diagram showing the various methods of cold, hot, and direct sample introduction.

molecular weight, may dehydrate in these circumstances, so that the corresponding olefin is also present in the vapour. Moreover, many acids, notably malonic and substituted malonic acids, cannot be examined in this way. None the less, the technique has proved immensely useful and even compounds in the two classes cited may be measured after chemical modification of the particular substance. The alcohols may be converted to the more volatile and much more stable silyl ethers (Sharkey Jr., Friedel and Langer, 1957) and the acids may be readily esterified to more stable materials.

There remains, however, further groups of compounds which are not amenable to these methods; also it may be of importance to the theoretical aspect of the origin of spectra that the cracking pattern of free acids or alcohols should be available. Sugar derivatives, too, which have a very low vapour pressure, require a different method of sample introduction, and there are many other natural products which suffer from the same limitation. To meet this difficulty, a technique of direct sample insertion is employed. The details of construction vary, but the essential principle is the same (Reed, 1958; Biemann, 1962 a, p. 28). The material is introduced into a small cylindrical tube (Fig. 7), preferably of uniform cross-section, which is positioned in proximity to the electron beam. The sample is also heated by some suitable means. The vapour in equilibrium with the sample in the tube is then usually sufficient to yield a spectrum. This technique has been successfully employed to obtain the spectra of furoic and malonic acids, sugar derivatives, many alkaloids, inositols, and even components of coal (Reed, 1960; Ryhage, 1960; Reed and Reid, 1963). Stable compounds of high melting point and low vapour

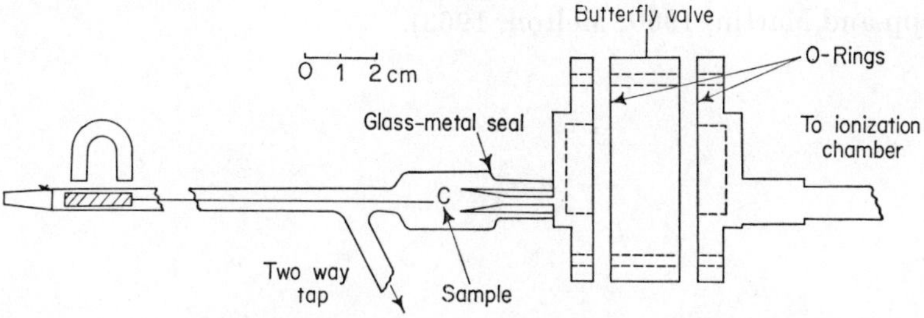

FIG. 7. Direct insertion system as used by Kelly and Reed. The butterfly valve allows isolation of the ionization chamber from the glass sleeve which can be opened to atmospheric pressure to change samples.

pressure, such as copper phthalocyanines (Beynon, Saunders and Williams, 1963; Hill and Reed, 1964), have also been successfully studied this way. The spectrum obtained may not, however, be a simple one, as the vapour pressure near the electron beam can be high, thus favouring ion–molecule collisions. These yield fragment ions which are additional to the normal spectrum (see p. 27).

Finally, in addition to the positive ion spectra obtained by the methods already discussed, there are negative ion spectra which are obtained by electron attachment.

$$M + e \rightarrow M^-$$

The negative ion so obtained may further dissociate to provide a series of negatively charged fragment ions, as well as neutral particles. In this way, a negative ion spectrum can be obtained which is characteristic of the compound examined. In principle, many of the instruments available for the production of positive ions may be simply modified to collect negative ions. However, the sensitivity of such machines is lower for negative than for positive ions. Also, the energy of the electrons in the beam must be rather closely controlled for successful attachment experiments, and it is better to employ a specially designed instrument for this form of analysis. One objection to an attachment method of obtaining spectra is that ions are observed having a mass greater than the original molecular ion and arising from the attachment of oxygen, hydroxyl, and other groups. Extensive fragmentation does not occur, and this reduces the information available in the analysis of an organic structure. However, the whole method of analysis provides interesting possibilities which have only recently been exploited, and it may become a useful supplementary method. The main publications in this field originate from one or two groups of workers only, and for further details reference should be made to their original communications (Ardenne and Tümmler, 1958; Ardenne, Steinfelder and Tümmler, 1961, 1962; Melton, Ropp and Martin, 1960; Melton, 1963).

CHAPTER 2

Concerning Mass Spectra

A. The Production of Spectra

The impact of electrons, usually of 50 eV or 70 eV energy, with a molecule yields a positive ion which may or may not dissociate to give rise to fragment ions. Polyatomic molecules can dissociate in many ways producing a spectrum rich in fragment ions:

$$e+\mathrm{ABC} \rightarrow \overset{+}{\mathrm{ABC}}+2e,$$

$$\overset{+}{\mathrm{ABC}} \rightarrow \overset{+}{\mathrm{AB}}+\mathrm{C}\cdot, \text{ or } \overset{+}{\mathrm{AC}}+\mathrm{B}\cdot, \overset{+}{\mathrm{BC}}+\mathrm{A}\cdot, \overset{+}{\mathrm{A}}+\mathrm{BC}, \text{ etc.}$$

The spectrum is reproducible provided the conditions of experiment are the same. This precaution extends not only to control of the electron bombardment energy and the gas pressure, but also to the design of the instrument itself. Thus, spectra which are run on instruments of different design, notably those which scan by varying the electrostatic potential, and those of a sector type which operate by varying the field strength of the magnet, produce similar, not identical, spectra for the same compound.

Table II compares the mass spectra of farnesol run on a Dempster, a time-of-flight and two sector-type instruments. The electron bombardment energy is the same in all cases. It can be seen that, apart from the base peak (which is common to only three spectra), the figures are merely in approximately the same order of abundance. There is not, as might be expected, a marked similarity in the spectra. Therefore, in order to establish the identity of compounds, the two spectra to be compared must be run under the same conditions in the same apparatus. Ultimately, it is likely that rules to correlate such variations may emerge, but apart from the generalization: "that the Dempster type instruments tend to discriminate against the heavier fragment ions", there has been only one detailed examination of the problem. This study, which has been carried out upon a series of hydrocarbons, has been reported by Caldercourt (1958), who concluded that the conversion of a complete cracking pattern from one instrument to a pattern obtained from another type of instrument cannot be performed satisfactorily.

TABLE II

Mass spectrum peaks for farnesol† obtained from various instruments

m/e	Time-of-flight	M.S.2	CEC	M.S.9
27	6·72	9·32	31·18	9·22
28		3·25	27·47	0·79
29	17·34	9·82	26·26	11·18
39	9·93	12·42	31·65	11·42
41	79·70	65·46	100·0	55·04
43	17·72	14·40	34·12	26·35
44	9·45	1·74	11·24	
51		2·56	12·54	2·70
53	10·48	11·21	31·52	9·62
55	18·70	23·76	49·10	14·70
67	18·27	18·12	21·39	11·57
68	12·83	8·07	10·45	12·20
69	100·0	100·0	82·75	100·0
70	7·62	6·92	18·52	7·22
77	5·38	11·12	18·84	3·05
79	9·65	21·02	23·60	7·22
81	34·56	23·12	20·89	21·81
91	7·46	15·54	18·70	5·57
93	24·19	61·16	59·61	20·48
94		10·11	10·69	4·02
95	11·68	8·74	8·73	8·28
105	4·72	11·01	12·74	3·59
107	10·15	20·18	17·64	7·96
109	9·27	10·77	10·24	7·73
119	5·32	15·68	18·43	4·98
121	6·91	9·82	11·31	5·94
133		16·79		
135		13·21		
161	3·02	11·21	10·24	2·88
189		2·56	2·50	1·01
191	0·82			1·07
203				
204			10·77	2·33
205			1·70	
206			0·31	
207				
220				
221				
222	1·32	0·41		0·82

†

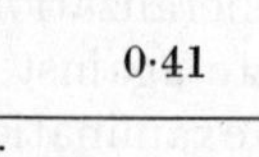

Incidentally, a similar discrimination against polynuclear substances has also been reported for electron multipliers as compared with the conventional detector.

Setting aside the peculiarities that arise from instrument design, the analysis of the mass spectrum obtained from a given compound is still a formidable problem and one, moreover, in which only limited successes have been achieved. The peculiar difficulty of the spectrum is that it is so rich in information of which, in the present state of the subject, we do not yet make the best use. For a given polyatomic molecule, excluding for the moment aromatic hydrocarbons, there is an ion in the spectrum corresponding to nearly every possible fragmentation. There are also ions which can only be obtained by re-arrangements of the initial molecular ion, and a series is often present of ions that result from bimolecular ion–molecule reactions. Further, the re-arrangement ions are usually molecular and they, too, may give rise to their own fragmentation pattern. In order to clarify the problem, it is intended to use the term "parent molecular ion" $(P)^+$ to indicate the compound introduced. Molecular ion $(M)^+$ will be used indiscriminately for any ion that may be formed as a subsequent step, whether by elimination with

$$C_4H_9\overset{+}{C}H_2CH_2Cl \rightarrow C_4H_9\overset{+}{C}H{=}CH_2 + HCl,$$

or without

$$C_5\overset{+}{H}_{11}CH_2COCH_3 \rightarrow C_5H_{10} + CH_3\overset{+}{C}OCH_3.$$

In cases where it is important to distinguish these processes, the acetone, as in the latter example, will be designated a "daughter molecular ion". The term "derived molecular ion" is not employed in this connection in order to avoid any confusion with derivative spectra (see Chapter 1, p. 5 *et seq.*).

Reference has already been made to the spectra produced by field ionization (see Chapter 1, p. 7) which are mainly composed of parent molecular ions. Such simplified spectra may also be obtained by electron bombardment at an electron energy which is only about 1 eV or less above the ionization potential of the molecule examined. The argument here being that as the average energy of the electron beam is only just greater than that needed to ionize the molecule, there is not enough energy also to break any bond in it. Figure 8 shows that the abundance of the parent molecular ion rises rapidly with increasing electron energy in the neighbourhood of the ionization potential, and it requires only to exceed this potential by a little to get a rather abundant molecular ion. This necessarily provides a much simpler mass spectrum, although it need not be entirely free of fragment ions.

The appearance potential of a fragment ion is related to its ionization potential (I) in the following way

$$A(\overset{+}{\mathrm{X}}) \geqslant I(\mathrm{X}\cdot) + D(\mathrm{X—Y}),$$

where A is the appearance potential of the ion and D(X—Y) is the bond dissociation energy of the bond binding X to the remainder of the

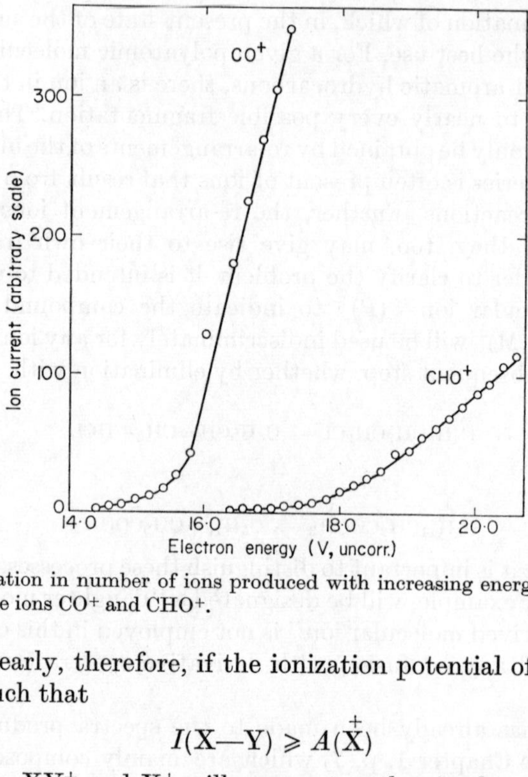

FIG. 8. Variation in number of ions produced with increasing energy of bombarding electrons for the ions CO^+ and CHO^+.

molecule. Clearly, therefore, if the ionization potential of the molecule (X—Y) is such that

$$I(\mathrm{X—Y}) \geqslant A(\overset{+}{\mathrm{X}})$$

then the ions XY^+ and X^+ will appear together in the spectrum. This phenomenon is not often observed, but sometimes the energy difference between parent and fragment ions is very small, as in the case of ethylamine (Hurzeler, Inghram and Morrison, 1958).

Because the ionization potentials of molecules differ somewhat over a range of a few volts and depend upon the structure, a knowledge of the potential and the molecular weight, which is of course obtained at the same time, is often of value (although of limited value) in analysing the structure.

Table III lists some molecules of varying complexity, together with their ionization potentials. Comparisons within a homologous series show that the potential decreases with increasing molecular weight. This decrease is at first marked, but eventually becomes negligible. Within a group of compounds of about the same mass but with different structure, it is generally true that branched compounds have a lower ionization potential than straight-chain ones. The introduction of one double bond facilitates ionization, as do heteroatoms possessing

TABLE III

Ionization potentials for the reaction

$$\mathrm{RH} \rightarrow \mathrm{R\overset{+}{H}} + e$$

Substance	Potential (eV)	Substance	Potential (eV)	Substance	Potential (eV)
Methane	13·12				
Ethane	11·65				
Butane	10·80	Isobutane	10·79		
Hexane	10·43	Isohexane	10·0	2-Methylpentane	10·3
Octane	10·24				
Decane	10·19				
Ethylene	10·56				
Propylene	9·80	Cyclopropane	10·23		
But-1-ene	9·72				
1,2-Pentadiene	9·42	Cyclopentene	9·27	Allene	10·16
1,3-Pentadiene	9·68			Methylacetylene	10·34
1,4-Pentadiene	9·58			Cyclopentadiene	8·9
2,3-Pentadiene	9·26				
1,5-Hexadiene	9·51				
Methyl	10·0	Cyclobutyl	7·88	Vinyl	9·45
Ethyl	8·78	Cyclopentyl	7·80	Allyl	8·16
n-Propyl	8·69	Cyclohexyl	7·66	Benzyl	7·76
Isopropyl	7·90			Methylcyclo-pentadienyl	8·54
n-Butyl	8·64				
Isobutyl	8·35				
s-Butyl	7·93				
t-Butyl	7·42				

lone-pair electrons. Aromatic compounds, too, have low ionization potentials. The value of these potentials in the analysis of pure compounds is limited by three factors. Firstly, apart from the lower members of a series, the differences in the potentials are so small that even quite precise measurements, which are often difficult, do not provide unambiguous solutions. Secondly, the sensitivity of most instruments is poor at low electron energies, and this requires a high sample pressure which is not always possible. Thirdly, the information that can be obtained is almost always available from the conventional spectrum.

In the case of mixture analysis, however, the method becomes particularly valuable, even with the use of double-focusing instruments, since it is easier to determine the molecular weight of individual components of the mixture from the simple low-energy spectrum.

A series of bond-dissociation energies is included in Table IV. It will be apparent that, compared with the energies normally employed in obtaining a mass spectrum, these are rather small. It is perhaps not surprising that, under normal operating conditions, such a variety of fragment ions do appear.

TABLE IV

Bond dissociation energy D (R—R′)

Substance	Energy (eV)	Substance	Energy (eV)	Substance	Energy (eV)
H—H	4·52	CH_3—CH_3	3·60	C_2H_5—H	4·16
H—OH	5·16	CH_3—C_2H_5	3·51	C_2H_5—NO_2	2·68
H—CN	4·93	C_2H_5—C_2H_5	3·38	C_2H_5—CHO	2·82
CH_3—H	4·38			C_2H_5—SH	3·00
CH_3—CN	4·64				
CH_3—OH	3·94			C_6H_5—H	4·42
CH_3—NH_2	3·47			$C_6H_5\cdot CH_2$—H	3·48
				$C_6H_5\cdot CH_2$—CH_3	2·74
				C_6H_5—C_2H_5	3·70

B. The Parent Molecular Ion

Ion bombardment nearly always yields a parent molecular ion, although the abundance varies widely. The presence of the ion, however, allows the molecular weight to be determined. Since the amount of material required for such a determination is, in favourable cases, of the order of 0·1 μg, this in itself constitutes a marked advance in analytical technique. Some compounds are known, particularly highly branched alkanes, which do not give a parent molecular ion; special methods are necessary for these. One convenient method is to run under rather high gas pressure, in order to produce ions from ion–molecule collisions, an approach which is examined subsequently (see p. 27).

Assuming that the parent molecular ion is present and in fair abundance, two methods are available to determine its molecular formula.

For single-focusing instruments of moderate resolving power, use is made of the isotope peaks. The method assumes that the distribution of the isotopes is that found in normally available material. Whilst enriched materials are valuable in the analysis of structure (see Chapter 5, p. 120 *et seq.*), they would not be useful in the present problem unless the degree

TABLE V

The mass spectrum of nitrobenzene (prominent ions)

m/e	% Abundance	Probable constitution	Calc. isotope contribution
30	15·82	$^{14}N^{16}O$	
31	0·28	$^{15}N^{16}O$	0·06
37	5·52	$^{12}C_3H$	
38	6·46	$^{12}C_3H_2 + {}^{12}C_2{}^{13}CH$	
39	9·97	$^{12}C_3H_3 + {}^{12}C_2{}^{13}CH$	
40	0·52		0·34
50	24·81	$^{12}C_4H_2$	
51	58·69	$^{12}C_4H_3$	
52	2·93	$^{12}C_3{}^{13}CH_3$	2·60
65	12·85	$^{12}C_5H_5$	
66	0·79	$^{12}C_4{}^{13}CH_5$	0·73
77	100·00	$^{12}C_6H_5$	
78	6·43	$^{12}C_5{}^{13}CH_5$	6·7
93	8·50	$^{12}C_6H_5{}^{16}O$	
123	42·27	$^{12}C_6H_5{}^{14}NO_2$	Parent
124	2·12	$^{12}C_5{}^{13}CH_5{}^{14}NO_2 + {}^{12}C_6H_5{}^{15}NO_2$	2·82
125	0·19	$^{12}C_4{}^{13}C_2H_5{}^{14}N^{16}O_2 +$ $+ {}^{12}C_5{}^{13}CH_5{}^{15}N^{16}O_2 +$ $+ {}^{12}C_6H_5{}^{14}N^{16}O^{18}O$	0·18
126	0·02	$^{12}C_5{}^{13}CH_5{}^{14}N^{16}O^{18}O$	

Since the normal abundance of any isotope is small, calculations have only been carried through for the next higher mass ion which follows a major peak. Two such peaks have been calculated in the case of the molecular ion. It will be noted that the observed isotope abundance is always slightly greater than that calculated, except for the isotopic ion above the parent molecular ion, $m/e = 123$.

of enrichment were known. A particular example, that of nitrobenzene (Table V), will make the process clear. The composition of the most abundant parent molecular ion is $^{12}C_6H_5{}^{14}N^{16}O_2$, corresponding to a molecular weight of 123. The natural abundance of carbon-13 is about 1·12%, i.e.

$$\frac{\text{No. of atoms of carbon-13}}{\text{Total No. of atoms}} \times 100 \simeq 1{\cdot}12\%.$$

The ratio of (carbon-13 × 100)/carbon-12 will be only very slightly greater ($\simeq 1{\cdot}13\%$). If the parent molecular ion is arbitrarily assigned an abundance of one hundred units, and the only isotopic contribution to the next higher peak $m/e = 124$ is that of the carbon-13, the abundance of this ion can be calculated as follows. Suppose that the molecule under investigation has only one carbon atom, then the abundance of this ion will be 1·13%. Nitrobenzene contains six carbon atoms. Each of these may contribute independently to $m/e = 124$. The total will be the product of the contribution from each atom times the number of such atoms, i.e. $6 \times 1{\cdot}13 = 6{\cdot}78\%$. The other three elements in nitrobenzene could also add to the abundance of the isotope peak $(P+1)^+$. Of these, however, only nitrogen-15 is of importance. The ratio $(^{15}N \times 100)/^{14}N = 0{\cdot}38\%$, and so this amount adds to the 6·78% arising from the carbon atoms. The percentage of naturally occurring deuterium is small (0·02%) and, unless there are many hydrogen atoms or an exact calculation is required, it may be neglected. In the present example the hydrogen isotope provides 0·1% abundance. The amount of oxygen-17 (0·04%) may be ignored. The calculated abundance of the ion $(P+1)^+$, $m/e = 124$ is about 7·3%.

The next higher mass, the ion $(P+2)^+$ ($m/e = 125$) may also be calculated. Any contribution from carbon-13 to the ion will depend upon the probability of one molecule possessing two such atoms, which is ${}^nC_2 \times 1{\cdot}13^2/100 = \frac{1}{200} \times 6 \times 5 \times 1{\cdot}28 = 0{\cdot}2\%$ for $P = 100\%$. Two oxygen-18 atoms taken singly provided 0·4% and the total abundance will be 0·6%. Therefore the ratios of $P:(P+1)^+:(P+2)^+$ are 100:7·3:0·6.

In practice this type of calculation may be reversed; and the relative abundance of the parent molecular ion and the further ions on the high mass side may be used to determine the chemical composition of the unknown. It is possible to set up a formula by which such an analysis can be conducted. If the abundance of the parent molecular ion P^+ is set at 100 and the compound possesses n carbon atoms, then the contribution from carbon-13 are $1{\cdot}13n$ for $(P+1)^+$ and ${}^nC_2 \times 1{\cdot}13^2/100$ for $(P+2)^+$. These are easily recognizable as the successive terms in a binomial expansion of the form

$$100(1+1{\cdot}13/100)^n = 100 + 1{\cdot}13n + {}^nC_2\,1{\cdot}13^2/100 + {}^nC_3\,1{\cdot}13^3/100^2 + \ldots$$

$$\overset{+}{P} \qquad (P\overset{+}{+}1) \qquad (P\overset{+}{+}2) \qquad (P\overset{+}{+}3)$$

It will be obvious that, unless the number of carbon atoms is very large or the material is enriched, the contribution to the appropriate ions becomes very small after $(P+2)^+$.

A corresponding calculation may be made for nitrogen. Here the ratio of 100 × No. of atoms nitrogen-15/No. of atoms nitrogen-14 ~ 0·38%, whence the successive terms may be obtained from the expansion $100(1+0{\cdot}38/100)^m$. Hydrogen can be calculated also.

The other commonly occurring atom is oxygen-18. The peculiarity here is that the isotope only adds to every other possible isotopic ion, namely P^+, $(P+2)^+$, $(P+4)^+$, etc. None the less, a similar calculation is possible.

All the expansions may be incorporated in one. For the molecule $C_kH_lO_mN_n$ the abundance of the ion P^+, $(P+1)^+$, $(P+2)^+$, etc., is given by the formula

$$\begin{aligned}100+(1{\cdot}13k+0{\cdot}02l+0{\cdot}38n)+0{\cdot}01({}^kC_2\times 1{\cdot}13^2+{}^lC_2\times\\ \times 0{\cdot}02^2+{}^nC_2\times 0{\cdot}38^2)+0{\cdot}2m+0{\cdot}0001({}^kC_3\times\\ \times 1{\cdot}13^3+{}^lC_3\times 0{\cdot}02^3+{}^nC_3\times 0{\cdot}38^3)+\text{etc.}\end{aligned}$$

which simplifies somewhat to

$$\begin{aligned}&\underset{\overset{+}{\mathrm{P}}}{100}+\underset{(\mathrm{P}\overset{+}{+}1)}{(1{\cdot}13k+0{\cdot}02l+0{\cdot}38n)}+\underset{(\mathrm{P}\overset{+}{+}2)}{0{\cdot}005(1{\cdot}28k(k-1)+0{\cdot}004\,l(l-1)+}\\ &+0{\cdot}14n(n-1))+0{\cdot}2m+0{\cdot}000017(1{\cdot}44k(k-1)\,(k-2)+\\ &\underset{(\mathrm{P}\overset{+}{+}3)}{+0{\cdot}000008\,l(l-1)\,(l-2)}+0{\cdot}044n(n-1\,(n-2))+\text{etc.}\end{aligned}$$

Among the other elements which may occur in organic compounds are the halogens, bromine and chlorine. The former possesses two stable isotopes, bromine-79 and bromine-81, which are equally abundant. The presence of a single bromine atom in a molecule is, therefore, easily recognized, for there are two parent molecular ions of equal abundance and two mass units apart. For bromobenzene, the two ions are at $m/e = 156$ and 158. The presence of two bromine atoms yields three ions, as with dibromodiphenyl, where the three possible species $C_{12}H_8{}^{79}Br_2$, $C_{12}H_8{}^{79}Br^{81}Br$, and $C_{12}H_8{}^{81}Br_2$ have parent molecular ions $m/e = 310$, 312 and 314, respectively. The abundances of the three ions are 50, 100 and 50, the most prominent parent molecular ion being arbitrarily set at 100.

In addition, of course, there will be the other isotope peaks which arise from the presence of carbon-13 in the molecule; these ions will appear at $m/e = 311$, 313 and 315. The abundance distribution at this end of the spectrum is given in Table VI. Substances containing three or more bromine atoms can be analysed in the same way. The treatment of organic compounds which contain chlorine atoms follows the argument given; the difference is that, since the relative abundances of the two naturally occurring isotopes chlorine-35 and chlorine-37 are nearly in the ratio of 3:1, the abundance of the parent molecular ions will differ accordingly. For the two ions $H^{35}Cl$ and $H^{37}Cl$, the molecular masses are 36 and 38 and the respective abundances are 100:33·3. Calculations for a compound containing two chlorine atoms show the ratio of abundances for the three parent molecular ions to be 9:6:1. Therefore *p*-dichlorobenzene will have three parent molecular ions of $m/e = 146$, 148 and 150 in the ratio of 100:67:11, respectively. The carbon isotope contributions occur at $m/e = 147$, 149 and 151 and amount to 6·8%, 4·5% and 0·7%, respectively. Again, this calculation may be extended to accommodate molecules with three or more chlorine atoms.

Consideration should be given to molecules which possess a mixture of chlorine and bromine atoms. Taking the nearly correct values of one and three for the ratios bromine-79/bromine-81 and chlorine-35/chlorine-37,

TABLE VI

m/e	Abundance	Ion species
310	50	$^{12}C_{12}H_8{}^{79}Br_2$
311	6·8	$^{12}C_{11}{}^{13}CH_8{}^{79}Br_2$
312	100	$^{12}C_{12}H_8{}^{79}Br^{81}Br$
313	13·6	$^{12}C_{11}{}^{13}CH_8{}^{79}Br^{81}Br$
314	50	$^{12}C_{12}H_8{}^{81}Br_2$
315	6·8	$^{12}C_{11}{}^{13}CH_8{}^{81}Br_2$

an abundance distribution can be worked out for substances containing both chlorine and bromine atoms. A formula can be set down, as for the more commonly occurring atoms discussed earlier. If the molecular species contains Br_a and Cl_b then the abundance distribution is obtained by multiplying out the formula $(^{79}Br + {}^{81}Br)^a \times (3^{35}Cl + {}^{37}Cl)^b$ and grouping together the terms which have the same mass. The general expression is tedious to work, but the following examples should make it clear. Consider a molecule that contains one bromine and one chlorine atom, i.e. $a = b = 1$. Then there is one parent molecular ion of highest mass containing the elements $^{81}Br^{37}Cl$, a second which is two mass units lower and made up of $^{79}Br^{37}Cl$ and $^{81}Br^{35}Cl$, and a third containing $^{79}Br^{35}Cl$, which is the parent molecular ion of lowest mass. The molecular weights of these ions are, therefore, $m+118$, $m+116$ and $m+114$ where m represents the non-halogen portion of the ion. The relative abundances are 1, 4 and 3, respectively. The value 4 is obtained as the ion $m+116$ is composite. It is made up of $^{81}Br^{35}Cl$ which, because $^{35}Cl/^{37}Cl = 3$, will be three times as abundant as $^{81}Br^{37}Cl$, and $^{79}Br^{37}Cl$ which will be as abundant as the latter.

As a further example, consider a substance of the formula $C_{14}H_7BrCl_2$. The molecular weights and ion abundances are given in Table VII. These figures are corrected for the presence of carbon-13 in the molecule.

Organic compounds may also contain sulphur. The element may be recognized by its isotopic distribution, in addition to the characteristic fragmentation of sulphur-containing molecules. Three isotopes, namely sulphur-32, sulphur-33 and sulphur-34 are those most commonly considered. The relative abundances are 95·1%, 0·7% and 4·2%, respectively. There is a further stable isotope, sulphur-36, but the abundance is so low, less than 0·02%, that it need not be considered. Once the presence

TABLE VII

m/e	Abundance	Ion species
324	59·6	$^{12}C_{14}H_7{}^{79}Br^{35}Cl_2$
325	9·4	$^{12}C_{13}{}^{13}CH_7{}^{79}Br^{35}Cl_2$
326	100·0	$^{12}C_{14}H_7{}^{79}Br^{35}Cl^{37}Cl + {}^{12}C_{14}H_7{}^{81}Br^{35}Cl_2$ $+ {}^{12}C_{12}{}^{13}C_2H_7{}^{79}Br^{35}Cl_2$
327	15·7	$^{12}C_{13}{}^{13}CH_7{}^{79}Br^{35}Cl^{37}Cl + {}^{12}C_{13}{}^{13}CH_7{}^{81}Br^{35}Cl_2$
328	47·6	$^{12}C_{14}H_7{}^{81}Br^{35}Cl^{37}Cl + {}^{12}C_{14}H_7{}^{79}Br^{37}Cl_2 +$ $+ {}^{12}C_{12}{}^{13}C_2H_7{}^{81}Br^{35}Cl_2$, etc.
329	7·3	$^{12}C_{13}{}^{13}CH_7{}^{81}Br^{35}Cl^{37}Cl + {}^{12}C_{13}{}^{13}CH_7{}^{79}Br^{37}Cl_2$
330	7·1	$^{12}C_{14}H_7{}^{81}Br^{37}Cl_2$
331	1·1	$^{12}C_{13}{}^{13}CH_7{}^{81}Br^{37}Cl_2$

of sulphur is suspected, the effect of these isotopes upon the abundance of the parent molecular ions may be calculated in the same manner as the other elements, previously discussed.

The determination of composition by isotope distribution is straightforward. There are, however, difficulties in the application of the method, some of which are so serious as to preclude its use. One of the less serious concerns molecules containing other atoms, such as fluorine, iodine, or phosphorus which, having only one isotope each, viz. fluorine-19, iodine-127, and phosphorus-31, may confuse the analysis unless their presence is suspected either by additional chemical information or from the cracking pattern. Of the three, iodine is perhaps the easiest to recognize because of its large mass; phosphorus may be the most difficult.

The absence of a parent molecular ion prevents the application of the method. Indirect methods are available such as that of effusion (Eden,

Burr and Pratt, 1951), but this can only be considered an approximation. Even when a molecular ion is present in low abundance, the analysis may be impossible, since the higher masses arising from isotopic elements may be immeasurably small or at least such that they cannot be measured with the required accuracy. For a molecule containing seven carbon atoms, the isotopic distribution resulting from the presence of carbon-13, $P^+:(P+1)^+:(P+2)^+$ is as 100:7·91:0·03. In the particular instance of 2,4-dimethylpentane, the abundance of the parent molecular ion $m/e = 100$ is 1·26% whence $(P+1)^+$ and $(P+2)^+$ become 0·01% and 0·003%, respectively, and far too small for accurate measurement.

A further weakness of the system, and one which is prevalent in the mass-spectrometry of organic molecules, is the protonation of organic ions. Under conditions of ion formation by electron bombardment, even some paraffinic hydrocarbons show a tendency to accept a further proton to give the ion RH_2^+, where RH represents the original hydrocarbon. It will be appreciated that this ion is of the same mass as the carbon isotope contribution in the simple mass spectrum of RH. As the normal isotope abundance is so small, a modest amount of the protonated ion will make analysis difficult. The error will always be in the sense of overestimating the number of carbon or nitrogen atoms present. It is prudent, therefore, to begin by employing this type of calculation to set an upper limit only to the number of such atoms. Because paraffins are so affected, olefins may be expected to show an enhanced effect which is further increased when compounds containing a hetero-atom with lone-pair electrons are present, e.g. sulphur, oxygen and nitrogen. Accordingly, many of the commonest organic substances (mercaptans, sulphides, alcohols, ethers, oxides, and amines) may provide intractable analytical problems. Some idea of the difficulties introduced by the basicity of the molecular ion is shown in Table VIII. The columns compare the observed and calculated abundances of the $(P+1)^+$ ion for the compounds named. The compounds are only roughly in order of increasing basicity, determined more by the intuitive chemical view than by the observed protonation. Thus, amines that readily form salts are considered more basic than ethers that only dissolve in strong mineral acids.

Double-focusing instruments are much better equipped to determine the composition of the molecular ion. Because the resolving power is so very high (as previously explained), mass defects become important, and measurements can be made precisely enough at least up to molecular weights of about 400, which either determine uniquely the chemical elements present or at least severely limit the number of alternatives. Since the measurement is made only upon the parent molecular ion, this need not be very abundant. Many compounds are now accessible to element analysis which, by reason of the small amount of the ion $(P)^+$, could not be examined previously.

The method supposes a series of tables in which the composition of all the possible alternatives together with the precise molecular weights are listed. Such tables are now available (Beynon and Williams, 1963); an interesting variant designed primarily to aid hydrocarbon analysis has been announced (Kendrick, 1963), and also an algorithmic method (Lederberg, 1964).

TABLE VIII

Molecule	% Abundance in excess of calculated	Molecule	% Excess in abundance of $\overset{+}{(P+1)}$
†Propane	0	Thiacyclobutane	0·78
Propene	0·03	2-Propanethiol	0·79
Ethyl methyl ether	0·07	n-Propylamine	0·90
Allene	0·10	Propionaldehyde	0·93
Propionic acid	0·10	2-Thiabutane	0·95
Cyclopropane	0·13	Allyl alcohol	0·98
1-Propanol	0·20		
Propyne	0·22		
Acetone	0·36		
Propylene oxide	0·51		
Methyl acetate	0·57		
Propane nitrile	0·31		
1-Propanethiol	0·04		

† Observed is based on a parent molecular ion abundance of 100. $(P+1)^+ = 3{\cdot}22\%$, which was taken as standard.

Too much reliance must not be placed upon these figures since it is known that the natural abundance of carbon-13 may vary somewhat. However, the observed variations are too great for this to be the whole explanation. Further, the discrepancy in general increases in compounds which one intuitively considers to be basic.

The mass spectrum which does not possess a parent molecular ion may no longer be impossible of solution. Double-focusing instruments determine the exact mass, and therefore elemental composition, even of fragment ions. Consequently, the mass determination of the heaviest ion in the spectrum can show a formula which is inconsistent with the hypothesis that it is the parent molecular ion, or even a molecular ion formed by rearrangement. As a particular example, one of the spectra discussed subsequently has ions $m/e = 113$ (2·45%) and $m/e = 114$ (0·23%). The heavier ion has the correct abundance to be the isotope contribution to $m/e = 113$. Exact mass measurement shows the latter to have the composition $C_8H_{17}^+$ and $m/e = 114$ has, therefore, the composition $^{12}C_7{}^{13}CH_{17}^+$. Now $C_8H_{17}^+$ is obviously derived from the radical C_8H_{17} and clearly the mass spectrum does not extend to the parent ion, which must have an even molecular weight, or contain an odd number of nitrogen atoms.

If the species possesses one or more hetero-atoms, detection is easier. Often there will exist a fragment ion containing atoms which are not present in the ion of largest mass. Alternatively, the smaller, abundant ion may possess more atoms of one element than are present in the higher mass. As an example, if the most abundant ion analyses for

TABLE IX

m/e	I Abundance	II Abundance	m/e	I Abundance	II Abundance
26	1·98		56	7·21	10·2
27	31·19	23·1	57	100	28·1
28	5·88	2·85	58	4·35	1·21
29	39·7	19·3	69	3·67	5·70
39	17·33	12·0	70	29·11	10·6
40	2·43	1·87	71	88·1	
41	44·15	31·1	72	4·67	
42	6·07	3·03	84		14·7
43	99·0	100	85		70·1
44	3·29	3·38	86		4·34
51	1·14		98	3·81	
53	3·89	2·40	99	15·54	
54	1·05		100	1·13	
55	20·49	7·20	113	3·61	2·39

$C_3H_6O^+$, whereas there is an abundant ion of constitution $C_5H_{11}^+$ also present, one correctly infers that these are two fragments of a more complex molecule.

While it is a difficult analysis, it is sometimes possible to deduce quite a considerable amount of the structure from the very fact that the parent molecular ion is absent. To anticipate the main discussion of a later chapter (see Chapter 3, p. 44), some idea of the procedure and its weaknesses is given. Table IX includes the fragment ions for two compounds. Analysis of the main ions indicates that they are all hydrocarbon ions; both substances are hydrocarbons of unknown molecular weight.

As a general observation, the fragmentation pattern of an alkane consists of two main series of ions of formula C_nH_{2n+1} and C_nH_{2n-1}. Moreover, the ion $m/e = 43$ or 57 is usually the base peak of the spectrum.

The ions of higher mass diminish progressively. The exception to this monotonic sequence arises when the hydrocarbon chain is branched. In these circumstances, an ion corresponding to the elimination of the longest chain is abnormally abundant, which may sometimes be accompanied by a re-arrangement ion of even mass. For a straight-chain alkane the parent molecular ion is usually (if occasionally only weakly) present.

Substance I has a base peak at $m/e = 57$ although $m/e = 43$ (99·0%) is only slightly less abundant. This implies that whatever the structure of the molecule, one branch probably consists of a C_4H_9 group attached to the centre of branching. The ion $m/e = 71$ is unusually abundant (89·0%), from which it may be inferred that another group in the molecule is $C_5H_{11}\cdot$; a conclusion which is supported by the presence of a re-arrangement ion $m/e = 70$. Further, the ion $m/e = 29$ (39·7%), which represents the ethyl ion, is very prominent. It is not clear how all these groups are attached to each other, or even if they are all present as distinct entities. Certainly it is unlikely that the ethyl or the propyl ion may be obtained by fragmentation of the butyl group present, or that the butyl may come from the amyl radical. However, it is possible that the ethyl or propyl may derive from the amyl. Allowing for these complications, one may write down several partial structures, including

$$
H_9C_4\text{—}CH\underset{\diagdown C_3H_7}{\overset{\diagup C_5H_{11}}{}} \quad , \quad H_9C_4\text{—}\overset{|}{\underset{|}{C}}\text{—} \quad \text{etc.}
$$
$$
\phantom{H_9C_4\text{—}C\text{—}xxxxx}C_2H_5
$$

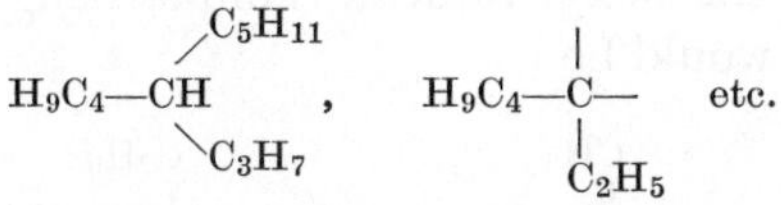

because, by the arguments advanced, one group must be butyl; the former may be rejected, since no ion corresponding to C_9 or C_{10} is present in the spectrum such as might be expected from this formula by the loss of one alkyl group. The second is possible, provided that the two valencies are satisfied in such a way that no fragment ion having more than seven carbons can be obtained. This follows, since the largest abundant ion $m/e = 99$ must have the formula $C_7H_{15}^+$, which limits the structure to the forms

$$
\underset{\overset{|}{C_2H_5}}{\overset{\overset{H}{|}}{H_9C_4\text{—}C\text{—}C_2H_5}} \quad \text{or} \quad \underset{\overset{|}{CH_3}}{\overset{\overset{CH_3}{|}}{H_9C_4\text{—}C\text{—}C_2H_5}}
$$

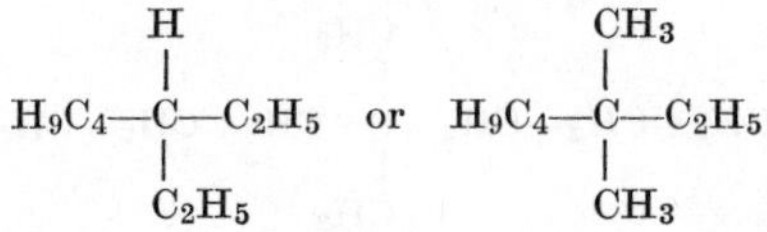

In the absence of any knowledge as to the effect of chain branching upon the abundance of the parent molecular ion, no further progress is possible. The correct molecular formula C_9H_{20} has, however, emerged.

Substance II has a similar type of spectrum with $m/e = 43$ as the base peak. The ion $m/e = 57$ $(28\cdot1\%)$ is not very abundant, but there is a re-arrangement ion associated with it $(m/e = 56)$ consistent with the $C_4H_9\cdot$ group being attached at a branch point in the chain. The ion $m/e = 85$ $(70\cdot1\%)$ is exceedingly abundant and it too is associated with a re-arrangement ion, $m/e = 84$. By analogy, therefore, we have two ions attached to a further branch in the chain. Again several possibilities exist, such as

$$\overset{\displaystyle C}{\underset{\displaystyle |}{\overset{\displaystyle |}{H_9C_4\!\!-\!\!C\!\!-\!\!C_6H_{13}}}}$$

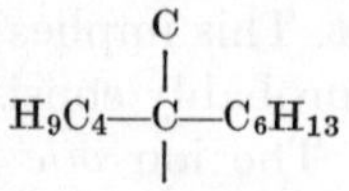

where the $C_3H_7\cdot$ group is attached by one of the unassigned valencies. Alternatively this group may be contained either in the $C_4H_9\cdot$ or $C_6H_{13}\cdot$ radicals. The first possibility may be rejected by reasons advanced for substance I; there is no evidence for a C_{11} fragment ion. The second can also be discarded for reasons already given, and therefore the three-carbon fragment must be contained within the six carbons. Further branching might exist in the $C_6H_{13}\cdot$ to provide a second point of easy fragmentation. No ion which could correspond to such a fission is observed. The above formulation is, therefore, wrong, and the C_4 and C_6 fragments are directly attached to each other, the branch carbon being contained in the hexyl radical. Summarizing all this evidence a plausible structure would be

$$\overset{\displaystyle CH_3}{\underset{\displaystyle CH_3}{\overset{\displaystyle |}{\underset{\displaystyle |}{H_9C_4\!\!-\!\!C\!\!-\!\!C_3H_7}}}} \quad \text{or} \quad \overset{\displaystyle C_2H_5}{\underset{\displaystyle H}{\overset{\displaystyle |}{\underset{\displaystyle |}{H_9C_4\!\!-\!\!C\!\!-\!\!C_3H_7}}}}$$

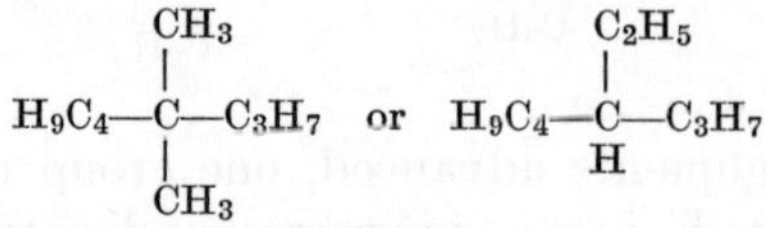

Both of these are incorrect. There is no piece of information available which correctly discloses the arrangement of the two fragments, viz. that the C_3H_7 and C_6H_{13} radicals are the two directly joined together. The branching occurs in the hexyl chain and there is no obvious way in which the formation of a butyl ion might be deduced. The substance is 4,4-dimethyl-n-heptane

$$\overset{\displaystyle CH_3}{\underset{\displaystyle CH_3}{\overset{\displaystyle |}{\underset{\displaystyle |}{CH_3\!\!-\!\!CH_2\!\!-\!\!CH_2\!\!-\!\!C\!\!-\!\!CH_2\!\!-\!\!CH_2\!\!-\!\!CH_3}}}}$$

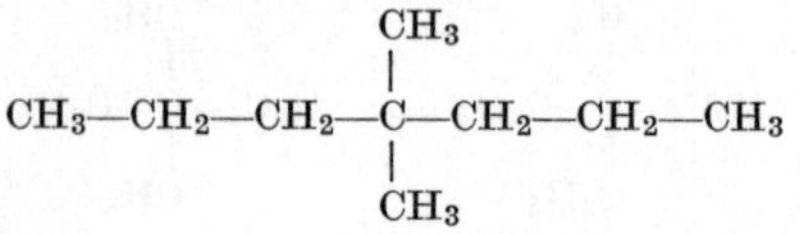

Comparable difficulties occur with compounds containing hetero-atoms although, in these, because of the way their presence modifies the fragmentation pattern, molecular structures can sometimes be identified.

C. Ion–Molecule Reactions

For any given compound, particularly if it is a basic molecule or the bombardment is carried out at rather high gas pressures, there is the likelihood of an ion–molecule collision reaction by means of which the parent molecule becomes protonated. The nature of such reactions and their importance in the spectra of organic compounds are discussed subsequently (see Chapter 5, p. 108). The mechanism of formation of these ions is not yet completely understood, although there is a growing volume of experimental evidence concerning them. The subject has been reviewed (Franklin, Field and Lampe, 1959; Melton, 1963) and reference may be made to this for a detailed discussion, as well as to Homer *et al.* (1963).

One possible origin of the commonest of such ions is by hydrogen extraction. If the parent molecular ion is RH^+ then the formation of the further ion RH_2^+ may be

$$\overset{+}{R}H + RH \rightarrow \overset{+}{R}H_2 + R\cdot$$

The method of formation is less important, for the present discussion, than a method of recognition. This is a simple matter since the ions are a product of a bimolecular process. For the simple mass spectrum, the extent of ionization of the substance is proportional to its partial pressure, provided that the operating conditions in the electron beam remain sensibly constant. If the ion RH_2^+ is formed in accordance with the above equation, its abundance will vary as the square of the partial pressure of RH. In many cases a plot of the abundance against pressure will suffice to detect ions of this class. However, many relatively non-volatile and even unstable compounds can be examined by the probe method (see Chapter 1, p. 9) where experimental pressure measurement may be impossible. Common sense often makes good the deficiency. From the nature of the method one may suppose a high concentration of vapour in the immediate neighbourhood of the electron beam. Thus, it is very likely that the measured spectrum contains some ions formed in the bimolecular collision process. In principle it may be possible to vary the vapour pressure of the compound by varying the temperature of the probe. The pressure varies in accordance with the relationship

$$\log_e p = -\frac{\Delta H}{RT} + \text{constant},$$

which is the integrated form of the Clausius–Clapeyron equation: where p is the partial pressure of the sample, T is the temperature on the absolute scale and ΔH is the latent heat of change.

This, however, is not often practical, as the temperature of the ion source is high, particularly in the region of the electron beam. Moreover, stringent control of the temperature of the vapour may be impossible.

Many examples, particularly the inositols, and furoic and malonic acids, which have been examined by the use of a probe, show unmistakable evidence of bimolecular collisions (Reed and Shannon, 1960; Reed and Reid, 1963). The phenomenon is very clearly shown in the inositols in particular, where there is a small ion corresponding to $(P+1)^+$. The parent molecular ion is absent. Further, when the composition of the molecule is known, isotope calculations as discussed in the previous section (see Chapter 2, p. 17 *et seq.*) will reveal discrepancies.

There remain a further group of compounds, e.g. proteins, to which none of these conditions apply, and which cannot be satisfactorily treated. For most organic problems, however, it is often sufficient to recognize the abundant fragment ions and to be able to give a molecular weight, requirements which can almost always be met.

Comparison studies which are being carried on seem to indicate that, with many categories of substance, the difference in the spectrum obtained by volatilizing the material in a hot box and upon a probe lie not so much in the risk of ion–molecule collision reactions, where the differences are often less than 1 %, but in the thermal degradation of the material evaporated in the heated tank (E. Clayton and R. I. Reed, unpublished observations). Some experiments with alkaloids and steroids have shown that the parent molecular ion is relatively more abundant when the probe is used than when the evaporated material is introduced into the ion source from a heated tank. The latter method gives a more abundant $(P-18)^+$ ion which could arise from thermal dehydration.

The double-focusing instrument may help with relatively simple molecular ions. High resolution might split the doublet made up of the RH_2^+ ion and the carbon-13 isotope ion of RH^+. The separation between these ions is rather small; they form a close doublet. Consequently, as the number of carbon atoms in the molecule increases, a very high resolving power is needed. An example will make this clear. Consider the ions of the doublet arising from the species $^{12}C_5H_{11}O^+$ and $^{12}C_4{}^{13}CH_{10}O^+$. The exact masses of these are 87·080985 and 87·076518, respectively. A measure of the resolving power needed for this doublet is given by $M_1+M_2/2\Delta M$ where M_1 and M_2 are the masses of the two ions and ΔM is the difference between them. Inserting the numerical values $87{\cdot}078752/0{\cdot}004467 = 19{,}493{\cdot}7$, a resolving power of 5 in 10^5 is needed.

The third possibility which has not yet been extensively examined is the nature of the derived spectrum. Theoretically, it can have a very high resolving power which can be controlled by the magnitude of the oscillatory component in the ion beam. Provided that every condition is ideal, it may prove possible to recognize the dissymmetry in the original ion beam from the difference between the two lobes of the derived spectrum. It does seem certain that the experimental difficulties could be formidable in the extreme.

One further, little used technique may be added for completeness. It has been shown by several workers that in simple cases the half-width of an ion beam, defined as in Fig. 9, may indicate the presence of excess kinetic energy associated with the ions.

As an example, half-beam width measurements, carried out upon molecular nitrogen and oxygen in the usual way, gave values which lie upon a rectilinear plot of $\delta P/P$. The value of P is chosen by adjusting

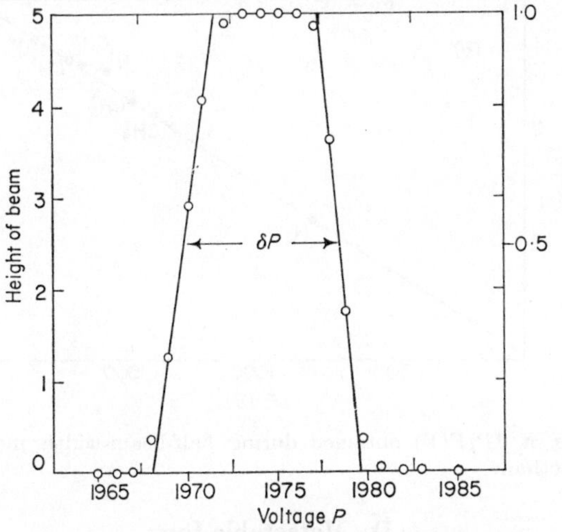

FIG. 9. Plot of beam height against voltage P.

the electrostatic potential and the field strength of the magnet, so that they are the same in all cases. The value of δP is then obtained by measuring the abundance of the ion under examination on varying the electrostatic potential at a fixed magnet strength.

The ions N_2^+ and O_2^+, together with H_2O^+, may reasonably be considered as normal reference ions. Similar measurements made upon N^+ ions, obtained as fragment ions in the spectrum of molecular nitrogen, gave a much larger value for the half-beam width and they lie above the rectilinear graph (Fig. 10). This indicated a broadening of the ion beam (McDowell and Warren, 1951). The method has never been extended to a study of even moderately complicated molecules, and it is by no means certain that valid deductions may be made from these. Certainly the experimental analysis will be difficult and tedious.

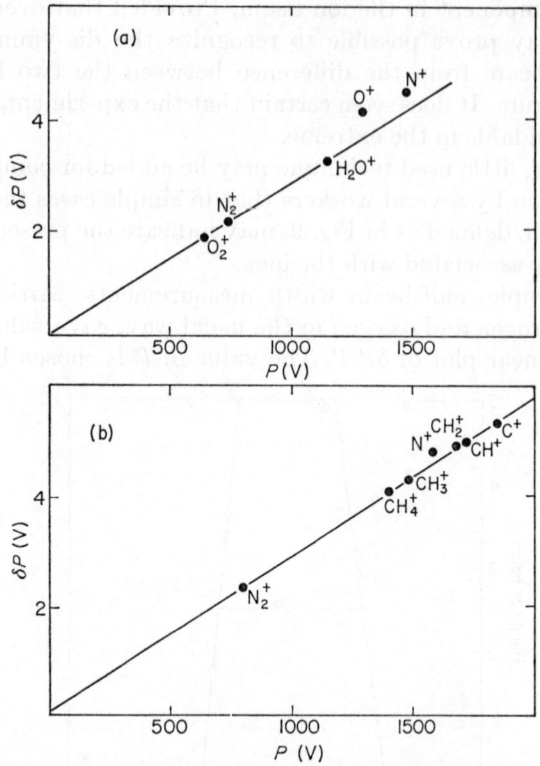

FIG. 10. Plots of $dP/P(V)$ obtained during half-beam-width measurements from (a) air and (b) methane.

D. Metastable Ions

Metastable ions originate in the dissociation of m_0^+ into m^+, an ion of lower mass m, between the original ion leaving the vicinity of the ion repeller plate and the newly formed fragment (or daughter) ion entering the magnetic field.

The ion is repelled electrostatically according to the equation

$$eV = \tfrac{1}{2}m_0 v^2$$

where V is the potential of the ion repeller, and v is the velocity of the ion. The daughter ion m^+ formed by fragmentation will possess $\frac{1}{2}(m/m_0)ev^2$ of the original energy. When the ion enters the magnetic analyser, the radius of gyration r_m is given by

$$r_m = \frac{1}{H}\{2V(m^2/m_0)\}^{1/2}$$

If one assumes that for these values of V and H there exists an ion m^{+*} the radius of gyration of which is equal to r_m, that is

$$r_{m^*} = r_m = \frac{1}{H}\frac{2Vm^{*1/2}}{e},$$

then

$$m^* = m^2/m_0.$$

Now, since m and m_0 are integers, m^* will in general not be. Metastable ions are rarely as well defined as normal ones and they are very much less abundant. These three characteristics make such ions readily recognizable.

The properties of metastable ions are now fairly well understood. The partitioning of the kinetic energy assumed in the derivation of the metastable ions has been verified for some (Hipple, Fox and Condon, 1946), although this may not be true in all cases. The metastable ion has been shown to have the same appearance potential as the ion from which it is derived (Fox and Langer, 1950), although some of these conclusions have been criticized (Field and Franklin, 1957). Further, the pressure of the metastable ion is proportional to the dependence of the species if the metastable ion is collision induced (Rosenstock and Melton, 1957). It is possible to examine the effect of slit width upon the abundance of this class of ions. As the resolution is increased by the narrowing of the slits, the abundance of the metastable ion is markedly diminished.

More recently, the metastable ions observed in the mass spectrum of methane have been further divided, the existence of collision-induced ions of this type having been demonstrated (Melton and Rosenstock, 1957).

The most important feature of metastable ions is that they allow one to determine the molecular mass, and sometimes the composition, of the neutral fragment elided, probably the only method by which such information may be obtained. Circumstance often allows a reasonable estimate of the elided fragment, as in the following instances. The molecular spectrum of 3-ethyl-n-hexane possesses a rather abundant ion $m/e = 85$. This corresponds to the loss of twenty-nine mass units which, inferentially, one assumes to be an ethyl radical. Indeed, such a presumption is necessary if one is to deduce the constitution of the molecule from the observed mass spectrum. It is possible, theoretically, that the net loss of this mass may refer to the elimination of two fragments from different parts of the molecule, viz. molecular hydrogen and a vinyl radical. In a similar way it is observed that the spectrum of *cis*-2-pentene has an abundant ion $m/e = 41$. In the latter spectrum, however, there is now present a metastable ion of $m/e = 24{\cdot}0$. Employing the formula developed previously, one can calculate the metastable ion referring to the process $70^+ \rightarrow 41^+ + 29$,

$$m/e = \frac{41^2}{70} = 24{\cdot}0,$$

which confirms the loss of an ethyl group as an entity,

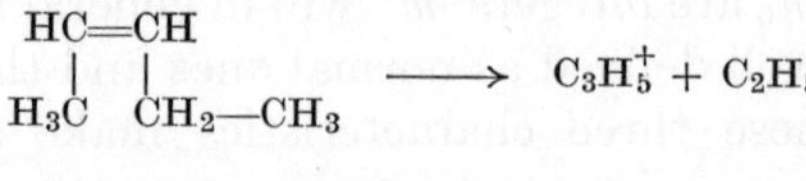

It is implied in the foregoing discussion that to obtain a metastable ion, fragmentation must occur after acceleration of the parent ion and before deflection of the daughter ion. This is clearly possible in sector instruments, in which there is a field-free space between the ionization source and the magnetic field. In such instruments, with which the author has had most of his experience, it has always been found that the observed value of the ion and that calculated from the probable fragmentation process have agreed (and usually well) inside 0·1 of a mass unit. In Dempster-type instruments, in which the ion source lies within the magnetic field, there is no field-free space where dissociation might occur. None the less, metastable ions are still obtained, but in lower abundance. In some instances, a divergence of as much as 0·4 mass unit has been claimed between observed and calculated values in such instruments (Biemann, 1962 a, p. 155).

It is easy to show that, in the case of sector as well as Dempster design instruments, the metastable ion occurs at a mass lower than either parent or daughter ion.

Consider the fragmentation $m_0^+ \rightarrow = m^+ + n$, where n is the neutral particle of mass n and m_0^+ and m^+ are as before. Now

$$m^* = \frac{m^2}{m_0} = \frac{m}{m_0}(m_0 - n) = m - \frac{mn}{m_0}$$

Since m and n are positive integers, $(mn/m_0) > 0$, even if fractional, whence it follows that $m > m^*$ and the metastable ion is of lower mass than the daughter ion. Obviously, the latter is less than the parent ion, which represents a formal proof of the theorem.

Certain instruments are available where the ions formed follow a cycloidal rather than a circular path. While a detailed discussion of the formulation of the equations for metastable ions in this instance is outside the scope of the present work, those formed in such instruments have two interesting properties. Firstly, the mass-to-charge ratio of the metastable ion may be greater than that of either the parent or daughter ion. Secondly, since the trajectory is not circular, the apparent mass of the metastable ion depends upon the point in the trajectory at which fragmentation occurs.

E. Other Considerations

As discussed in the preceding pages, electron bombardment of an organic material in the vapour phase leads to the production of ions. These usually include the parent molecular ion and a series of fragment ions. The whole assemblage of ions constitutes the cracking pattern. Provided that the operating conditions are the same, this pattern will be repeated for the same compound (in the same instrument). A comparison of the spectra obtained in two instruments of the same design and operated in the same way shows that, while the two spectra are not identical, they resemble each other closely enough to demonstrate that they are both of the same substance. The identity of the operating conditions is important even in trying to reproduce a spectrum exactly in the same instrument. This may be due to several causes. Firstly, the production of ions is sensitive to many factors, notably the potential surfaces in the source. Secondly, variations in the sample pressure result in changes in the contributions made by ion–molecule reactions.

Differences in design often cause significant changes in the cracking pattern, a subject previously mentioned (p. 11).

The most valuable application of mass spectrometry as a tool in organic chemistry would be to take the observed spectrum and to deduce the original molecular structure from it. It is also, however, very useful to be able to reject a possible structure merely by a consideration of the cracking pattern.

The initial problem is either to treat the cracking pattern upon a qualitative basis and investigate the structural pattern from first principles, or to compare the unknown with other compounds that may prove relevant to the structure problem in the same instrument. Both types of argument may be employed. It is considered most important, however, that the supposed identity of the unknown be checked by obtaining the mass spectrum of an authentic sample under identical conditions. The identity of the two spectra will be good evidence for the correctness of the deductions.

The average time which the ion remains in the source is about 10^{-5} sec. The ion formed may undergo a unimolecular reaction for which the frequency factor will be of the order of 10^{13} sec^{-1}. Thus, very many vibrations may occur in the ion during its residence in the source. Provided that it possesses the necessary activation energy, the original ion may fragment or re-arrange during this time. Experimental evidence indicates that the average activation energy of such reactions of ions is small, usually of the order of 0·2–0·7 eV (Field and Franklin, 1957).

It is reasonable to assume, having regard to the 70 eV energy of the incident beam, that the ions are possessed of at least this small amount of excess energy. Thus, a large variety of fragment ions may be expected in the final spectrum, an expectation which is generally realized.

Further properties that are relevant are the bond dissociation energy and the ionization potential of the fragment ion. As stated, the ion AB^+ may fragment to yield either $A^+ + B\cdot$ or $A\cdot + B^+$. In, for example, the fragmentation of n-nonane, the cleavage which yields $C_5H_{11}^+ + C_3H_7\cdot$ will provide the complementary pair of fragments, namely $C_3H_7^+$ and $C_5H_{11}\cdot$.

The energies required to produce these are

$$A(C_5\overset{+}{H}_{11}) \geqslant I(\cdot C_5H_{11}) + D(H_{11}C_5—C_3H_7)$$

for the amyl ion, and for the propyl

$$A(C_3\overset{+}{H}_7) \geqslant I(\cdot C_3H_7) + D(H_{11}C_5—C_3H_7)$$

The bond dissociation energy will be the same in each instance, and therefore the factor which leads to the preferred formation of one or other ion is the difference

$$A(\overset{+}{C}_5H_{11}) - A(\overset{+}{C}_3H_7) = I(\cdot C_5H_{11}) - I(\cdot C_3H_7)$$

making the further plausible, but untested, assumption that the excess energy associated with the ionization of each radical is nearly the same for both (cf. Stevenson, 1951), for $I(\cdot C_5H_{11}) \simeq I(\cdot C_3H_7) \simeq 8{\cdot}6$ eV and either ionization may occur. Therefore the mass spectrum will contain both ions.

On the other hand, if $A = \cdot C_3H_7$ and $B = \cdot CH_3$, the molecule being n-butane, then the probability of one or other ionization process occurring will be determined by the difference

$$I(C_3H_7\cdot) \simeq 8{\cdot}6 < I(\cdot CH_3) = 10{\cdot}0 \text{ eV}$$

Thus, there is very much less likelihood of methyl ions appearing in the mass spectrum and, except in special circumstances (Chapter 3, p. 49), it is rarely abundant in mass spectra.

Since the time allowed for decomposition of the parent molecular ion is great (by comparison with the frequency factor for unimolecular decomposition reactions), and the energy requirement is low, one may expect some evidence in the observed spectrum for nearly every possible fragment ion. This is so, and the average spectrum is bewilderingly rich in detail. In any case it would make it difficult to pick out the fragmentations which are significant for the determination of molecular structure. In reality, the problem is even more complex, since there are present re-arrangement ions also. Some will be fragment ions derived after re-arrangement in the molecular ion; others will be further fragment or re-arranged fragment ions of these, and yet others will be daughter molecular ions and possibly fragment ions pertaining to them. Although such complexities often utterly confuse the analyst and prevent any extensive deductions, nevertheless, provided that they can be recognized, re-arranged ions may at least provide some identification of the class of compound examined.

A classical example of a re-arrangement which is very confusing is the appearance of the fragment ion $m/e = 29$ (38·5%) in the spectrum of neopentane. This requires a preliminary re-arrangement, or at least one which is concurrent with decomposition

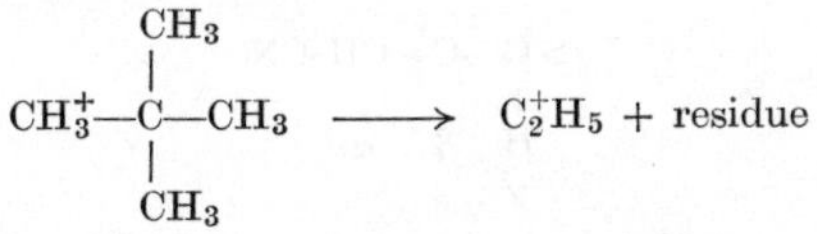

Now there is abundant evidence from kinetic investigations of neopentyl derivatives that when the neopentyl cation is formed it readily re-arranges to the isoamyl ion. This does not seem to be true in the present study. The abundance of the neopentane ion is zero in the cracking pattern; that of the isopentane ion in its own spectrum is 6·19%. It does not seem possible, therefore, that neopentane first isomerizes to isopentane before the fragmentation which yields the ethyl ion; at best it can be a concerted process. Recent evidence has shown that the central carbon of neopentane is present to a large extent in the ethyl ion (38%) (Rylander and Meyerson, 1956). The investigation of the mechanism of such re-arrangements is irrelevant to the present study and those interested should consult the original publications.

Re-arrangements of the second kind referred to above may be helpful in the analysis. The spectrum of capronitrile shows an ion $m/e = 41$. This is the acetonitrilium ion which is formed by the cleavage of the cyanomethylene ion from the parent molecular ion with the concomitant removal of a hydrogen atom

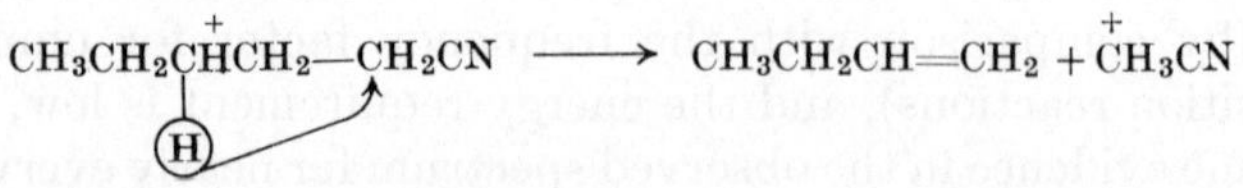

It is, of course, not the only fragmentation that occurs, but the ion so produced is prominent in the overall spectrum. The but-1-ene ion, which may also be formed, may further decompose as possibly does the acetonitrile ion. The spectrum might therefore contain, in addition to the fragment ions of capronitrile, those of the two molecular ions formed by re-arrangement.

The importance of recognizing this rather common process of fission accompanied by re-arrangement is that it gives us two important indications of the structure of the original material. Firstly, the grouping $-CH_2CN$ is present, otherwise acetonitrile could not be formed. Secondly, the γ-carbon must possess at least one hydrogen atom for hydrogen migration to occur.

The information so obtained is enough in this simple illustrative example to give considerable insight into the molecular structure. Combining the two pieces of analytical deduction, one can write a partial structure of the form

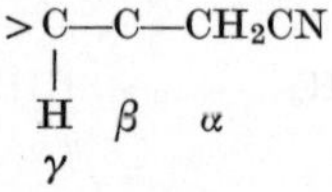

which, together with the molecular weight obtained from the rather abundant parent molecular ion, leaves only two carbon and eight hydrogen atoms to assign. Their distribution may be decided on the basis of other fragment ions.

The spectrum has been obtained on a double-focusing mass spectrometer, and the fragment ions are shown together with their abundances and probable composition in Table X.

The formation of the alk-1-ene is conventionally supposed to arise in the re-arrangement process. So far, no definite evidence has been obtained which proves the correctness of this assumption, although it is clearly the most plausible. Again, the problem is not very important to the present study.

The abundance of the parent molecular ion is of considerable significance in analysis. An abundant ion indicates great stability and, as a consequence, the fragmentation pattern is sparse and may be feeble. Compounds which have an extensive conjugation often belong to this category. The electron removed in the ionization is almost certainly one

TABLE X

High resolution spectrum of capronitrile

m/e	Species	Abundance (%)	m/e	Species	Abundance (%)
27	HCN^+	47·8	64		1·59
28			65		0·9
29	$C_2H_5^+$	55·3	66		2·1
30	$C_2H_6^+$	1·8	67		2·5
31	Impurity $CH_3NH_2^+$	1·0	68	$C_4H_6N^+$	36·2
36		0·5	69	$C_5H_9^+$, $C_4H_7N^+$	27·5
37		2·6	70		4·5
38		5·0	71	$C_5H_{11}^+$	0·9
39	$C_3H_3^+$	29·9			1·1
40		6·2	75		0·2
41	CH_2CN^+	75·2	76		0·3
42	CH_3CN^+	13·9	77		0·4
43	$C_3H_7^+$	25·7	78		0·4
44	$C_3H_8^+$	2·2	79		1·5
45	Impurity $C_2H_5NH_2^+$	1·2	80		1·1
49		0·4	81		1·5
50		2·2	82	$C_4H_8CN^+$	26·3
51		3·3	83		2·3
52		5·3	84		0·3
53		6·1	85		0·4
54	$C_2H_4CN^+$	100·0	91		0·6
55	$^{12}C_2{}^{13}CH_4N^+$, etc.	34·8	92		0·4
56	$C_2H_5CN^+$	5·3	93		0·5
57	$C_4H_9^+$	39·6	94		0·4
58	$C_4H_{10}^+$	2·2	95	$C_5H_9CN^+$	1·1
59		0·4	96	$C_5H_{10}CN^+$	16·9
61		0·3	97	$C_5H_{11}CN^+$	1·8†
62		0·4	98		0·3
63		0·7			

† Parent molecular ion.

of the π-electrons contained in the double bond. Delocalization of the π-electrons results in a distribution of the positive charge formed over all the conjugated system.

Amongst the hydrocarbons, n-alkanes have an abundant parent molecular ion that makes a diminishing contribution to the total spectrum with increasing molecular weight. Olefins are, in general, even more stable than the corresponding saturated hydrocarbon. Cyclic compounds, too, have a more abundant parent molecular ion than the corresponding n-alkane and often more than the olefin of the same molecular weight. The customary argument advanced for this is that in order to produce fragment ions two bonds have to be broken instead of one. This is certainly so, but may not be the whole reason.

Amongst the alkane series the presence of isomers has a marked effect

TABLE XI

The abundances of the parent molecular ions of isomeric nonanes

Isomer	Abundance (%)	Isomer	Abundance (%)
n-Nonane	6·38	3-Methyl-3-ethylhexane	0·02
2-Methyloctane	2·08	3-Methyl-4-ethylhexane	1·51
3-Methyloctane	1·65	2,2,3-Trimethylhexane	0·03
4-Methyloctane	2·46	2,2,4-Trimethylhexane	0·03
3-Ethylheptane	1·17	2,2,5-Trimethylhexane	0·05
4-Ethylheptane	1·27	2,3,3-Trimethylhexane	—
2,2-Dimethylheptane	0·06	2,3,4-Trimethylhexane	0·19
2,3-Dimethylheptane	1·03	2,3,5-Trimethylhexane	1·49
2,4-Dimethylheptane	0·66	2,4,4-Trimethylhexane	—
2,5-Dimethylheptane	1·07	3,4,4-Trimethylhexane	0·03
2,6-Dimethylheptane	2·80	3,3-Diethylpentane	—
3,3-Dimethylheptane	—	2,2-Dimethyl-3-ethylpentane	—
3,4-Dimethylheptane	1·28	2,3-Dimethyl-3-ethylpentane	—
3,5-Dimethylheptane	0·65	2,4-Dimethyl-3-ethylpentane	0·16
4,4-Dimethylheptane	—	2,2,3,3-Tetramethylpentane	—
2-Methyl-3-ethylhexane	0·69	2,2,3,4-Tetramethylpentane	0·02
2-Methyl-4-ethylhexane	0·89	2,2,4,4-Tetramethylpentane	0·02
		2,3,3,4-Tetramethylpentane	—

upon the abundance of the parent ion. A secondary alkane has a less prominent molecular ion than the corresponding primary one whereas, when a tertiary centre is present, the ion is insignificant or entirely wanting.

Individual exceptions may occur, but these generalizations are a good guide to the probable structure. Table XI compares the abundance of parent molecular ions amongst a series of isomers, and these conform to the principle that the more branching present the lower the abundance, as discussed above.

The introduction of hetero-atoms further lowers the stability of the parent ion, an effect which is conveniently summarized in Table XII.

Strict comparisons are not possible, since the differing valencies of the hetero-atoms mean that with some, e.g. nitrogen, a centre of branching

TABLE XII

Compound	Abundance of P^+ (%)	Compound	Abundance of P^+ (%)
n-Propane	39·5	‡1-Bromopropane	13·7
Ethanol	100	†2-Chloropropane	23·6
Ethyl mercaptan	100	Propionitrile	10·0
Ethylamine	18·8	Ethyl methyl ether	25·8
Isobutane	3·9	Thiabutane	63·5
2-Propanone	13·1	n-Propanol	37·4
n-Butane	16·9	n-Pentane	8·7
2-Fluoropropane	1·2	Diethyl ether	30·4
†1-Chloropropane	3·0	n-Butanol	44·8
‡2-Bromopropane	10·6	3-Thiapentane	69·4

† $C_3H_7{}^{35}Cl$ considered as the parent molecular ion.
‡ $C_3H_7{}^{81}Br$.

is introduced simultaneously. The problem may be avoided to some extent by comparing cyclic structures (Table XIII).

Certain general observations are, however, possible. The introduction

TABLE XIII

Compound	Abundance of P^+ (%)	Compound	Abundance of P^+ (%)
Cyclopropane	100	1,4-Dioxane	20·6
Ethylene oxide	65·2	Cyclopentylmethanol	100
Cyclopentane	29·3	α-Tetrahydrofurfuryl alcohol	0·21
Thiacyclopentane	52·1	Methylcyclopentane	15·9
Tetrahydrofuran	0·1	3-Methylcyclopentanone	67·4
Pyrrolidine	20·4	1,3-*cis*-Dimethylcyclopentane	14·4
Cyclohexane	74·9		
Thiacyclohexane	100		
Piperidine	100		

of a singly bonded hetero-atom always lowers the stability of the parent ion. The apparent exception is the introduction of sulphur. The opinion has been advanced that the apparently small variation caused by introducing this atom arises from the rough cancellation of two opposing effects (Biemann, 1963). Firstly, the introduction of the hetero-atom lowers the stability of the ion. Secondly, the sulphur gives the molecule a much greater ion cross-section, thereby increasing the probability of ion formation. Such an effect may be seen in the comparison of the cross-sections obtained for methane, water and hydrogen sulphide using 75 eV electrons (Lampe, Franklin and Field, 1957). While these simple molecules are far from being as complex as most organic compounds, they nevertheless illustrate the effect. The appropriate values are $Q = 4{\cdot}65$, $2{\cdot}96$ and $6{\cdot}42$; where $Q \times 10^{-16}$ cm^2 is the ionization cross-section. There is a sharp fall on replacing CH_2 by oxygen, which is more than compensated if sulphur is substituted instead.

The general comparison is a very rough one, but does show that the introduction of hetero-atoms with lone pair electrons weakens the structure. The effect is often obscured in cyclic compounds, as any strain which is present in the carbocyclic system may be relaxed by the introduction of a hetero-atom such as sulphur or, to a lesser extent, oxygen; the bond angle is more easily deformed in the heterocycle.

CHAPTER 3

The Fragmentation of Hydrocarbons

While it should be emphasized that a successful analysis of a fragmentation pattern must take into account the distribution of, and mutual interactions amongst, all the atoms in the molecule (indeed the limited number of examples already given show the importance of atom distribution particularly in respect of ion abundance and re-arrangement processes), it is none the less convenient arbitrarily to dissect the molecular structure. For the purposes of examination, therefore, four classes of structural element will be discussed separately. Their synthesis into a single molecule is the basis of subsequent chapters (see Chapter 5, Section A *et seq.*).

An organic molecule may be regarded as a carbon skeleton to which are attached hetero-atoms or groups containing these. Such groups may be further divided. Firstly, there are those which are terminal, as a cyano group is terminal in capronitrile, or the hydroxyl group in n-heptan-3-ol (i.e. the oxygen atom is not included in the carbon skeleton). Secondly, there is that class in which the hetero-atom is an integral part of the skeleton: these include the furans, di-n-propyl sulphide, piperidine, and so on.

In addition, there are of course compounds possessing more than one functional group; a very large class which may be further differentiated if required. One such subdivision would include compounds having two or more functional groups, e.g. glycerol, ethyl aminoacetate, epichlorhydrin, etc. A second would comprise mixtures of the previous categories, such as in ethyl furoate, β-methoxypyridine and physostigmine. The third would contain two hetero-atoms either in separate systems as with ketomanoyl oxide, or in the same ring as with *N*-methylmorpholine.

The amount of information available concerning the various classes differs widely. The most exhaustively surveyed, thanks to the immense effort of the Petroleum Industry, is undoubtedly the hydrocarbons. At the other end of the scale are the polyfunctional molecules, where the information available has been the result of the use of mass spectroscopy as an adjunct to other methods of structure analysis. It is convenient for the present purpose to examine the classified groups in the following

Glycerol

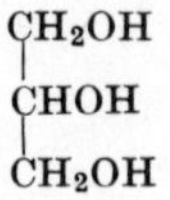

Physostigmine

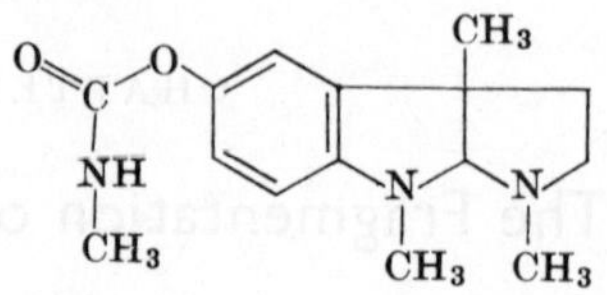

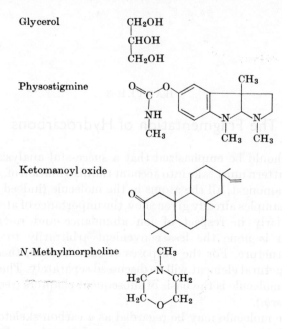

Ketomanoyl oxide

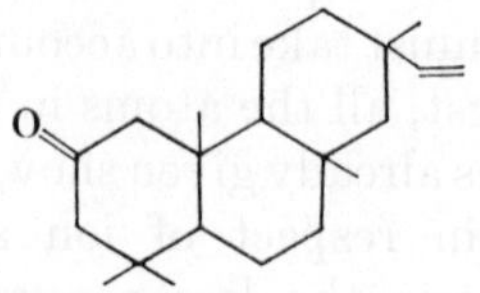

N-Methylmorpholine

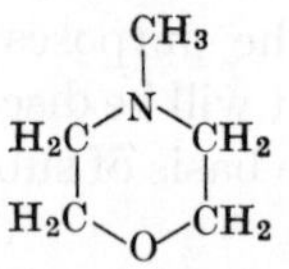

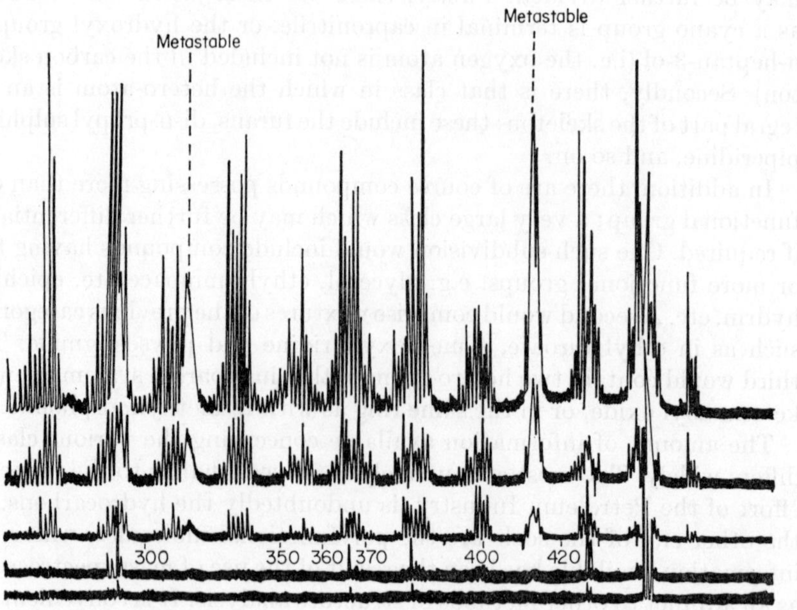

Fig. 11. Mass spectrum showing presence of metastable peaks.

order: I, the hydrocarbons; II, the attached groups containing hetero-atoms; III, the skeletal system which contains a hetero-atom; IV, the polyfunctional molecules. It will be useful, moreover, to include under II certain groups, such as methoxyl, which may be placed logically under III and, conversely, to consider the carbocyclic group and acid amide groups as single entities rather than to deal with them under IV. The apparent irregularities in the classification will be considered under II.

A. The Hydrocarbons

Several classes of hydrocarbon can be distinguished and, in order to deal with them systematically, they will be further subdivided into:

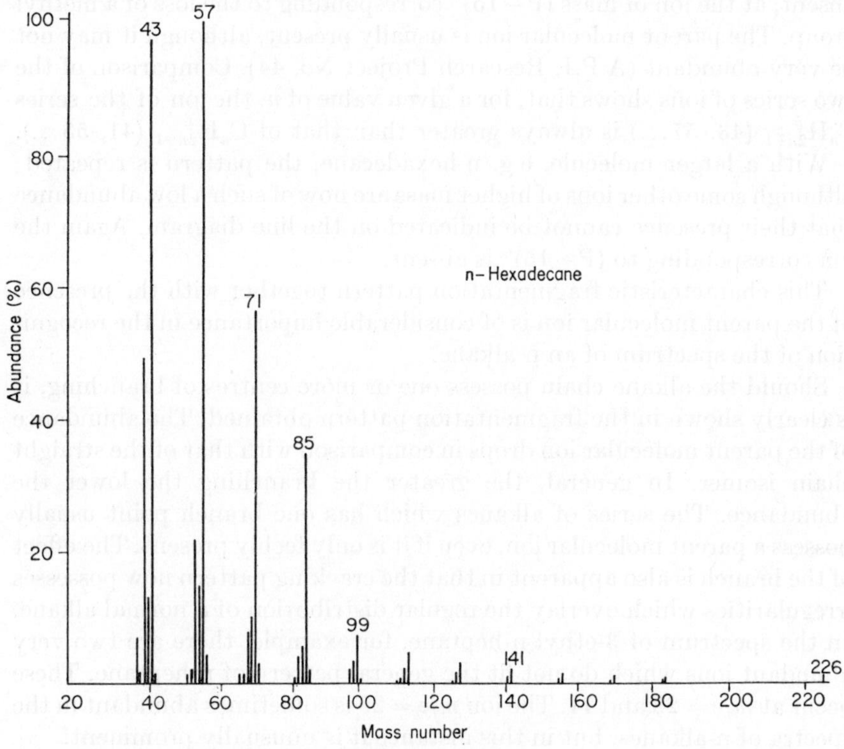

FIG. 12. n-Hexadecane.

1, alkanes; 2, cycloalkanes; 3, bicyclo- and polycycloalkanes; 4, alkenes; 5, cycloalkenes; 6, dialkenes; 7, alkynes; 8, aromatics; 9, miscellaneous structures.

The alkanes are probably the most extensively investigated of any group of organic compounds. In the straight chain series the general pattern of fragmentation is as shown in Fig. 12.

The spectrum of relatively small molecules shows two series of ions, formulae $C_nH^+_{2n+1}$ (43, 57, 71...) and $C_nH^+_{2n-1}$ (41, 55, 69...) respectively. Most ions which do not belong to these, with the exception of the parent molecular ion, are usually insignificant. Both sets reach their maximum value in this instance when $n = 3$. It must be noted in passing that Dempster type instruments almost always give the ions of maximum abundance to be $m/e = 41$ and 43, respectively. Sector type instruments frequently give $m/e = 55$ and 57, respectively as the major ions in these series. As one moves to the higher homologues, the abundance of the ions in both series diminishes; becoming very small, or more frequently absent, at the ion of mass $(P-15)^+$ corresponding to the loss of a methyl group. The parent molecular ion is usually present, although it may not be very abundant (A.P.I. Research Project No. 44). Comparison of the two series of ions shows that, for a given value of n, the ion of the series $C_nH^+_{2n+1}$ (43, 57...) is always greater than that of $C_nH^+_{2n-1}$ (41, 55...).

With a larger molecule, e.g. n-hexadecane, the pattern is repeated; although some other ions of higher mass are now of such a low abundance that their presence cannot be indicated on the line diagram. Again the ion corresponding to $(P-15)^+$ is absent.

This characteristic fragmentation pattern together with the presence of the parent molecular ion is of considerable importance in the recognition of the spectrum of an n-alkane.

Should the alkane chain possess one or more centres of branching, it is clearly shown in the fragmentation pattern obtained. The abundance of the parent molecular ion drops in comparison with that of the straight chain isomer. In general, the greater the branching the lower the abundance. The series of alkanes which has one branch point usually possess a parent molecular ion, even if it is only feebly present. The effect of the branch is also apparent in that the cracking pattern now possesses irregularities which overlay the regular distribution of a normal alkane. In the spectrum of 3-ethyl-n-heptane, for example, there are two very abundant ions which do not fit the general pattern of n-heptane. These occur at $m/e = 29$ and 71. The ion $m/e = 29$ is sometimes abundant in the spectra of n-alkanes, but in this instance it is unusually prominent.

The abnormal abundance of the ions is in accordance with the observation that branched hydrocarbons fragment preferentially at centres of branching, with the preferred elimination of the longer branch.

There is in addition a re-arrangement ion of even mass $m/e = 70$ which results from a hydrogen elimination that is concurrent with fission to yield the ion $C_5H_{10}^+$. It has been pointed out (Friedman and Long, 1953) and now well established that in the fragmentation of hydrocarbons the odd mass ions are always more abundant than those of even mass. When,

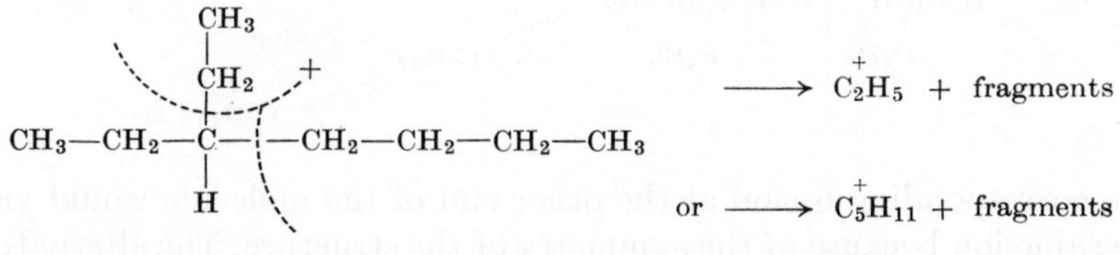

therefore, an even mass ion is observed this is a useful indication of the occurrence of a re-arrangement.

It is interesting to note that if one branch in the chain is a methyl group there is now nearly always some evidence for its loss. The $(P-15)^+$ is present in contradistinction to its absence from the spectra of the n-alkanes.

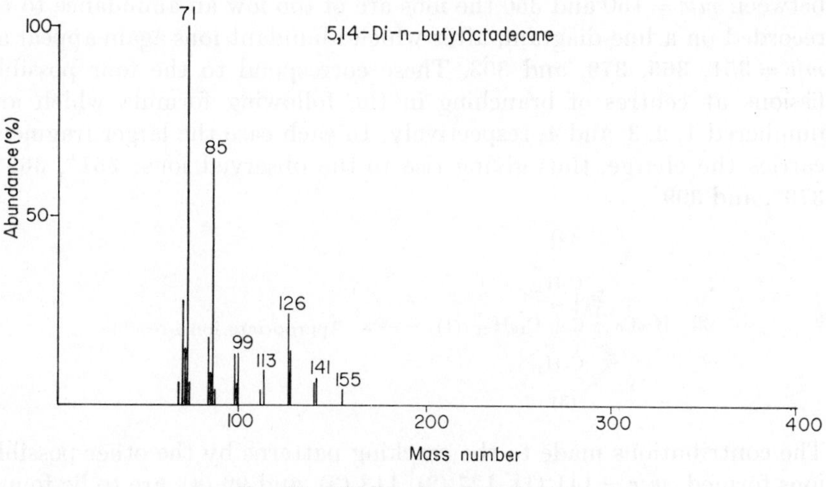

FIG. 13. 5,14-Di-n-butyloctadecane.

The introduction of a second branch into the hydrocarbon, but on a different carbon atom, again lowers the stability and, hence, the abundance of the parent molecular ion. It is now rather small and, in compounds of higher molecular weight, absent all together. Figure 13 shows the cracking pattern of 5,14-di-n-butyloctadecane, a compound which incidentally does not possess a parent molecular ion. The fragmentation characteristic of an alkane occurs at the low mass end of the spectrum. The observed irregularity in the abundances of $m/e = 126$ and

127, which are too prominent in comparison with their neighbours, is associated with the preferred fragmentation at a branch.

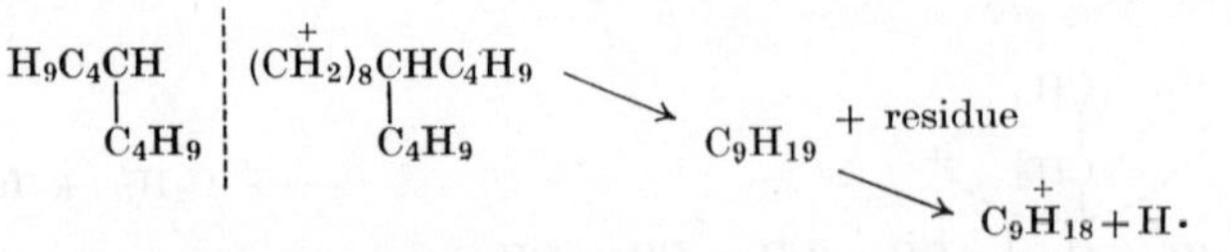

The corresponding fission at the other end of the molecule would yield the same ion because of the symmetry of the structure. The alternative, namely that formed by the elimination of butyl and having $m/e = 309$ ($C_{22}H_{45}^+$), is also present. Its abundance, since it represents over 10% of the base peak, is too large in regard to its position in the spectrum to be an n-alkane.

In alkanes with a quaternary centre, the parent molecular ion is always absent. Such a compound is 10-n-heptyl-10-n-octyleicosane. The low mass end of the spectrum again shows the usual distribution; between $m/e = 150$ and 350 the ions are of too low an abundance to be recorded on a line diagram, after which abundant ions again appear at $m/e = 351$, 365, 379, and 393. These correspond to the four possible fissions at centres of branching in the following formula which are numbered 1, 2, 3, and 4, respectively. In each case the larger fragment carries the charge, thus giving rise to the observed ions: 351^+, 365^+, 379^+, and 399^+.

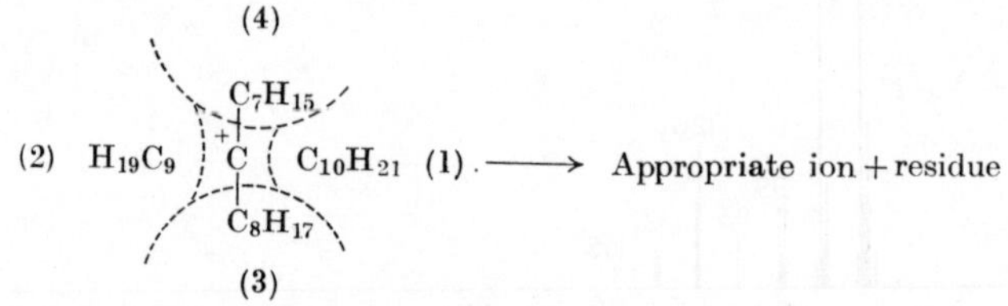

The contributions made to the cracking patterns by the other possible ions formed, $m/e = 141$ (1), 127 (2), 113 (3), and 99 (4), are to be found amongst those at the low-mass end of the spectrum.

This property of the parent molecular ion, namely that it fragments at a centre of chain branching with elimination of the largest branch, is often very useful in determining the structure of a hydrocarbon.

The spectrum of 3-ethylhexane is shown in Table XIV. The parent molecular ion $m/e = 114$ together with the high abundance of the isotope peak ($m/e = 115$) proves the constitution of the molecule to be C_8H_{18}. Three items in the mass spectrum show that we are dealing with a branched alkane.

(i) The abundance of the parent molecular ion is only 1·6%, whereas that of n-octane is 6·7% of the base peak.

(ii) The ion $m/e = 85$ is unusually abundant with reference to the remainder of the spectrum; as also is the ion $m/e = 71$.

TABLE XIV

Spectrum of 3-ethylhexane

m/e	Abundance (%)	*m/e*	Abundance (%)	*m/e*	Abundance (%)
1	0·23	43	100·	71	12·6
2	0·36	44	3·35	72	0·67
12	0·02	45	0·04	73	0·02
13	0·05	49	0·02	74	0·02
14	0·21	50	0·28	75	0·01
15	2·29	51	0·72	77	0·09
15·3	0·02	52	0·30	78	0·02
16	0·08	53	1·93	79	0·06
21·7	0·15	54	0·76	80	0·01
25·1	0·06	55	11·2	81	0·04
26	1·29	56	5·99	82	0·01
26·0	0·21	57	12·4	83	0·18
27	21·9	58	0·53	84	22·8
28	2·89	59	0·01	85	28·7
29	19·9	61	0·01	86	1·78
30	0·43	61·9	0·03	87	0·05
37	0·21	62	0·11	91	0·01
37·1	0·06	63	0·09	93	0·01
38	0·43	64	0·03	98	0·03
38·2	0·11	65	0·23	99	0·07
39	9·98	66	0·09	100	0·01
39·2	0·36	67	0·43	101	0·02
40	1·40	68	0·11	113	0·02
41	22·6	69	2·51	114	1·63
42	6·69	70	12·1	115	0·14

(iii) There is an abundant ion of even mass (70^+) present in the spectrum; and since the molecule is a hydrocarbon, the ion must result from a re-arrangement.

From this information one concludes that the unknown is a branched chain alkane. Since one abundant ion corresponds to the ion $(P-29)^+$

an ethyl group is present. This leaves five carbon atoms and thirteen hydrogens to assign, for a partial structure must be of the form

$$-\overset{\displaystyle |}{\underset{\displaystyle C_2H_5}{C}}-$$

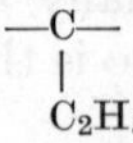

As there is a parent ion (P^+) albeit a small one, it may be concluded that the molecule does not possess a quaternary carbon atom. Accordingly, one part of the compound is

$$\overset{H}{\underset{C_2H_5}{>C<}}$$

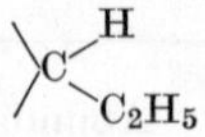

The ion $m/e = 70$ corresponds to $C_5H_{10}^+$ which is a re-arrangement ion and most probably refers to breakage at the point of chain branching as with 3-ethyl-n-heptane. Therefore one part of the skeleton comprises a chain of five carbon atoms.

$$C-C-\overset{H}{\underset{\diagdown}{C}}-$$
$$CH_2$$
$$\diagdown$$
$$CH_3$$

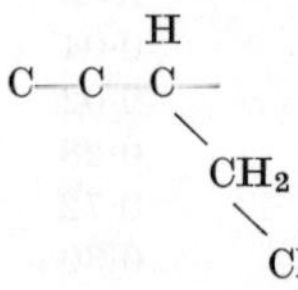

Three carbon atoms remain to be assigned. The ion $m/e = 43$ is very abundant which allows a three-carbon branch in the original structure. For the same reason, the relatively weak ion $m/e = 57$ is consistent with the view that there is not a four-carbon chain attached to the centre of branching. This reduces the possible number of isomeric octanes to two, either

$$\underset{\displaystyle CH_2-CH_3}{\overset{\displaystyle H}{CH_3CH_2CH_2-\overset{|}{\underset{|}{C}}-CH_2CH_3}} \quad \text{or} \quad \underset{\displaystyle CH_3 \; CH_2-CH_3}{\overset{\displaystyle H}{CH_3CH_2-\overset{|}{\underset{|}{C}}-CH_2CH_3}}$$

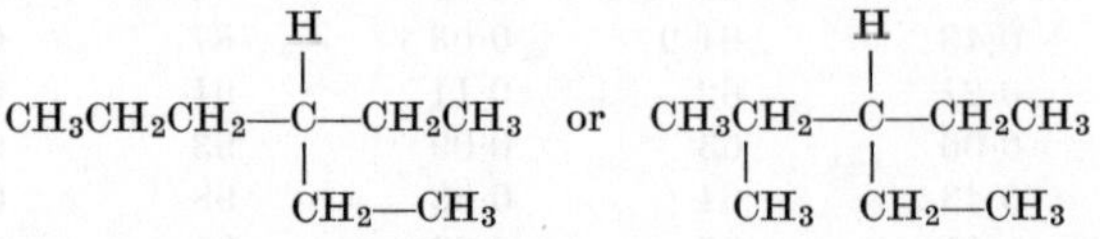

A final decision between them is not easy. However, the bulk of the available experimental evidence suggests that when there is both an abundant $m/e = 43$ and $m/e = 41$, the group is most probably n-propyl. When, on the other hand, the ion $m/e = 41$ is rather small compared with $m/e = 43$, the original group was probably an isopropyl. In the present example 43^+ (100%) is very much more abundant than 41^+ (23%). Accordingly one might deduce, incorrectly, that the C_3H_7 is an isopropyl and that the original molecule was 2-methyl-3-ethylpentane instead of 3-ethyl-n-hexane, the correct solution.

The spectrum of 2-methyl-3-ethylpentane is known. In this, the relative abundances of the ions 43^+ and 41^+ are as 26·4% to 100%; no greater disparity than in the one analysed. The analysis clearly demonstrates the difficulties encountered in attempting to deduce even a rather simple structure from the relative abundances of the ions formed.

B. Cycloalkanes

The cyclic structures possess a different cracking pattern from the alkanes. The parent molecular ion is more stable than in the corresponding alkane, as is shown in Table XV.

TABLE XV

Spectrum of cyclopentane

m/e	Abundance (%)	*m/e*	Abundance (%)	*m/e*	Abundance (%)
12	0·03	37	0·85	57	0·05
13	0·06	38	2·61	61	0·15
14	0·38	39	21·50	62	0·33
15	3·01	40	7·24	63	0·47
16	0·07	41	29·3	64	0·07
25	0·06	42	100·0	65	0·67
26	2·85	43	3·26	66	0·30
27	14·60	49	0·07	67	1·54
28	3·96	50	0·57	68	0·41
29	4·62	51	0·92	69	1·07
30	0·11	52	0·29	70	29·5
31	0·28	53	1·93	71	1·53
32	0·28	54	0·92	72	0·02
33	0·19	55	28·8		
36	0·01	56	1·23		

Among the simpler ring systems it is apparent that the abundant fragment ions are $m/e = 83$, 55, 41, and 27. These belong to the series $C_nH_{2n-1}^+$ which is consistent with a molecular formula of C_nH_{2n}. The ion $m/e = 83$ is either the base peak or very abundant in a series of mass spectra of compounds of the form R—C_6H_{11} where R = Me, Et, n-Pr, n-Bu, n-octyl, n-decyl and 2-cyclohexyloctane, as shown in Table XVI. Since the formula of the abundant ion is C_6H_{11} it is plausible to regard this as originating in fission at the branch carbon with the preferential formation of the cyclohexyl ion. This is true even of methylcyclohexane, in contrast to branched alkanes where, although the loss of a methyl group is apparent, the cleavage seldom yields the base peak of the spectrum. The alkyl radical rarely appears as an ion in the spectra of cycloalkanes. This may be the result of the alkyl radical having a much higher

TABLE XVI

Spectrum of 2-cyclohexyloctane

m/e	Abundance (%)	m/e	Abundance (%)	m/e	Abundance (%)
29	42·0	68	4·83	110	6·96
30	0·90	69	41·0	111	33·0
36.7	0·38	70	17·19	112	24·7
37	0·06	71	25·5	113	3·68
37·1	0·21	72	1·30	114	0·21
38	0·29	73	0·02	115	0·02
39	17·48	77	1·14	123	0·10
39·2	0·17	77·3	0·04	124	0·19
40	3·28	78	0·26	125	0·83
41	77·1	79	2·21	126	0·35
42	19·04	80	0·54	127	0·08
43	42·1	81	6·28	137	0·02
44	1·31	82	100·0	138	0·08
50	0·19	83	93·9	139	0·26
51	0·93	84	11·49	140	0·05
52	0·69	85	1·29	141	0·02
53	7·53	86	0·06	152	0·07
54	10·33	91	0·40	153	0·20
55	76·4	92	0·07	154	0·04
56	18·63	93	0·20	166	0·10
57	40·1	94	0·07	167	0·47
58	1·67	95	1·06	168	0·08
59	0·03	96	0·29	178	0·03
61·8	0·08	97	4·41	179	0·05
62	0·01	98	0·52	180	0·04
63	0·13	99	0·10	181	0·46
63·4	0·09	103	0·02	182	0·07
64	0·04	105	0·05	195	0·02
65	1·11	107	0·05	196	4·70
66	0·91	108	0·02	197	0·71
67	32·6	109	0·54	198	0·05

ionization potential than cyclohexyl. It is more likely, however, that the radical further decomposes. The presence of the abundant ion $m/e = 83$ is useful in that it indicates a monosubstituted cyclohexane. Since the parent molecular ion is also present to provide the molecular weight in compounds of this class, there is usually little difficulty in their identification. Even so, there is a further indication that a ring system is present, namely the rather abundant $m/e = 39$ being the ion $C_3H_3^+$ (Breslaw and Gal, 1959).

Another indication of the essential correctness of the view that the main fragmentation is the preferred elimination of R· comes from an examination of the cracking pattern of doubly substituted cyclohexanes. A series of dimethyl derivatives together with 1-ethyl-1-methylcyclohexane and 1,2,3-trimethylcyclohexane have been examined. Even those with quaternary carbons have a parent molecular ion to assist in molecular weight determinations. The ion which corresponds to the elimination

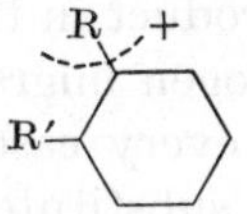

to yield the ion $R'—C_6H_{10}^+$ is still abundant in most cases, but it is no longer the base peak in every such compound. The base peak is sometimes the ion $m/e = 55$, presumably the result of fission across the ring

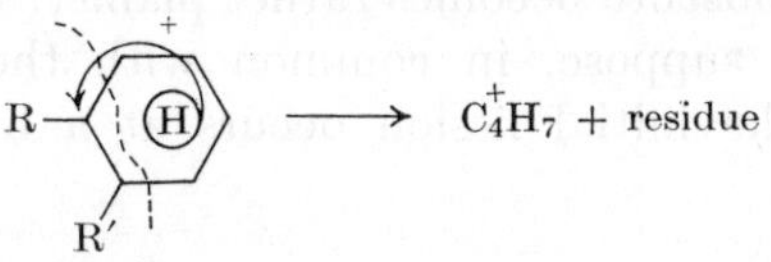

plus a concomitant hydrogen re-arrangement. The ion 39^+ is also abundant.

A further problem is presented in these substances since the two substituent groups, if on different carbon atoms, may be disposed *cis* or *trans* to one another. The relative spatial distribution which these two groups occupy will depend upon the conformation adopted by the cyclohexane ring. Much work remains to be done in this field, although valuable contributions have already been made (Natalis, 1963). This author's view is that while the ring is flexible there will be only minor differences in the fragmentation patterns. It is true that there are differences in energy levels between boat and chair conformations of cyclohexane, but these differences are relatively small compared with the gross energies available in impact experiments. Moreover, on the formation of the molecular ion there is formed also an electron deficient bond. Whether such a link will allow further distortion of the conformation is as yet undecided. In any event, all factors of this kind tend to minimize the differences observed. In the case of rigid or nearly rigid systems, however, the small changes in the abundance of fragment ions become more marked, as mentioned subsequently (see Chapter 5, E, F). Further, in the one case reported here where the two substituents are different, namely 1-ethyl-1-methylcyclohexane, preferred elimination of the larger, the ethyl, group occurs. This is quite consistent with the previously reported fissions in branched alkanes.

There remain several obscure points for which detailed study will certainly be needed. One such concerns the base peaks of the spectra of both 1,4-*cis*- and *trans*-dimethylcyclohexane, which occur at $m/e = 55$. The simple fission mentioned above is now inadequate as an explanation. One must suppose either fission along the path necessary to yield the correct fragment, or some re-arrangement.

As mentioned in an earlier chapter (see Chapter 2, E), much of the writing of alk-1-enes as one product in the fragmentation of molecular ions with a concomitant hydrogen migration is conventional; decisive evidence is lacking in nearly every case. Equally, it is by no means certain that the fission of the substituted cyclic molecular ions is not accompanied by a re-arrangement. The probability of a cleavage across two bonds occurring at the same instant is small. A sequential fragmentation is preferred. If this is accompanied by re-arrangement much that was formerly obscure becomes rather plainer. Thus, in the present example one may suppose, in common with the bulk of reported experience, that the initial fission occurs at a branch point in the molecular ion (P^+).

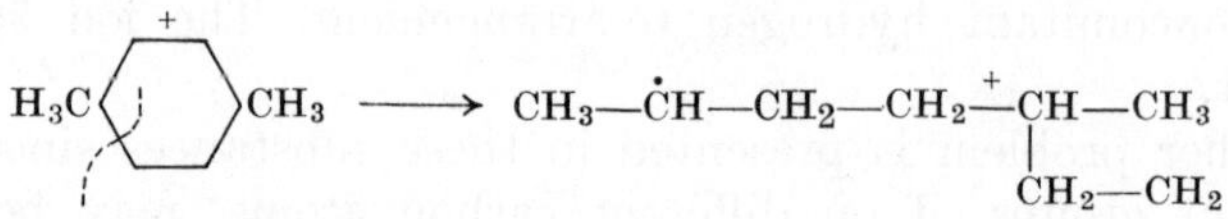

With the rupture of this bond, all distinction between the *cis* and *trans* disposed systems is lost; free-rotation is now possible. As a further postulate, this breaking is accompanied by a hydrogen migration to yield an alkene ion, i.e.

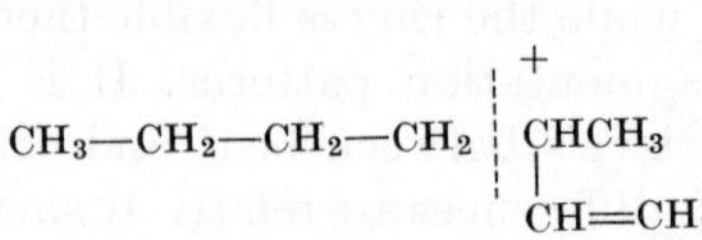

To anticipate the discussion of the next section, it will be noted that, in accordance with general observation, such ions break preferentially at an allylic bond. Applying this consideration, the above ion must yield $m/e = 55$ ($C_4H_7^+$).

It is not considered that this demonstrates the existence of an equilibrium between the cyclic and acyclic structures. Possibly there exists some intermediate state, between the cyclic and acyclic ion, when the migrating hydrogen is associated with the whole molecular ion rather than bound to a particular carbon atom that would allow the loss of the identity of isomers which depend upon molecular geometry without implying that the migration is irreversible. As a further complication, there is the likelihood that the hydrogen migration and the breaking of the second bond may be concerted processes.

While little systematic evidence of the correctness of this postulate exists, it is none the less a useful hypothesis in practical analytical spectrometry and has been used successfully in this laboratory.

Similar fragmentations occur in mono-substituted cyclopentanes, although naturally the mass of the fragment ion resulting from the loss of the alkyl group is $m/e = 69$ ($C_5H_9^+$) rather than $m/e = 83$. It is interesting that the other frequently abundant ion is still 55^+, in agreement with the proposed re-arrangement.

C. Alkenes

A second series of compounds, the alkenes, have the same general formula C_nH_{2n}. In this series, the parent molecular ion is more abundant than in the corresponding alkane. It approximates closely to that of the equivalent cycloalkane. The fragmentation pattern is distinct from the alkanes but again resembles the cycloalkanes in having the more abundant ion in a C_nH_{2n-1} series (55, 69...).

The interpretation of the cracking pattern in a straight chain alkene is a relatively simple matter; the most facile fissions occur at the allylic bond in the molecular ion. This has long been known (Brown and Gillams, 1954) and is of great use in interpretation. It is a common experience in conventional chemistry that the allylic bond dissociation energy is below that of the average carbon–carbon bond, and is distinctly weaker than a vinylic bond. The ionization potential of the fragment containing the allyl group will be below that of an alkyl group, in general and, therefore, by the following relationships:

$$A\,(\mathrm{R\,CH\overset{+}{=}CH{-}CH_2}) \geqslant I\,(\mathrm{R\,CH{=}CH{-}CH_2\cdot}) + D\,(\mathrm{R\,CH{=}CH{-}CH_2{-}\vdots{-}R'})$$

where R and R′ are alkyl groups. Now $I\,(\mathrm{R\,CH{=}CH{-}CH_2}) \ngtr I\,(\mathrm{R'\cdot})$ and, therefore, $A\,(\mathrm{R\,CH{=}CH{-}CH_2^+}) < A\,(\mathrm{R'^+})$ (Chapter 2, A) where the symbols have their usual significance. Accordingly, the fragment formed by fission of the allylic bond will be very prominent in the spectrum. If the alkene has a high molecular weight, some evidence of fragmentation of the alkyl chains may be present.

The spectrum of non-4-ene (Fig. 14) shows a characteristic progression corresponding to the $C_nH_{2n-1}^+$ series. The ion $m/e = 97$ is rather abundant considering the absence of higher mass ions and therefore one can infer that there is an allylic bond between the seventh and eighth carbon atoms, resulting in the $C_7H_{13}^+$ ion. A moderately abundant parent molecular ion determines the molecular formula as C_9H_{18} and hence the structure is R—CH═CH—CH_2—C_2H_5 where R is C_4H_9. A corresponding analysis would reveal the presence of the allylic bond from the ion

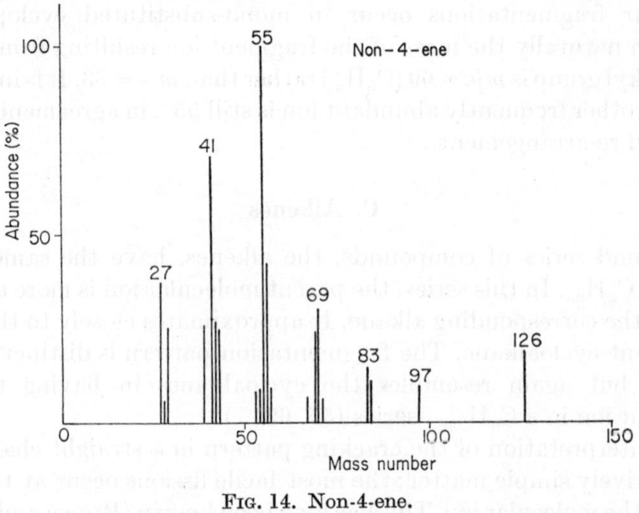

FIG. 14. Non-4-ene.

83^+ ($C_6H_{11}^+$). In order to detect that this ion is in fact unusually abundant, considerable experience in the overall cracking patterns of straight chain alkenes is needed. Accepting this information in the absence of such a detailed examination, one deduces that the original molecule is of the form C_3H_7—CH_2—CH═CH—CH_2—C_2H_5, and it only remains to decide whether the end group is a normal or isopropyl radical.

Applying the tentative argument previously advanced, that the occurrence of an abundant ion (43^+) indicates the presence of an isopropyl group, one might conclude that, because the more abundant ion is 41^+, the group is an n-propyl, which leads in this instance to the correct structure (non-4-ene). It is clear that arguments of this kind are less convincing than those employed for alkanes.

The analysis of the branched alkene 4-ethyloct-2-ene proceeds as follows. The molecular ion $m/e = 140$ ($C_{10}H_{20}^{+}$, 10%) is present, confirming the molecular weight. The base peak of the spectrum is at $m/e = 55$ which, together with the molecular formula of $C_{10}H_{20}$, confirms that the molecule is indeed an alkene; we have excluded cycloalkanes from the present discussion. By comparison with the spectra of other alkenes, it is apparent that two fragment ions, namely $m/e = 98$ and 111, are unusually abundant; the masses correspond to the ions $C_7H_{14}^{+}$ and $C_8H_{15}^{+}$. The second, an alkenyl ion, represents the loss of twenty-nine mass units from P^{+}. Assuming an allylic fission, one then obtains a partial structure

$$\mathrm{R{-}CH{=}\underbrace{C{-}CH}_{H,\,R'}{-}C_2H_5}$$

where $R + R' = C_5H_{12}$. The other fragment, of even mass, is a re-arrangement ion. Experience in the cracking patterns of alkanes shows that this is probably related to fragmentation at a branch carbon. Therefore, the C_7 group is most probably attached at a tertiary carbon. The complete molecule comprises only ten carbon atoms, consequently some of the atoms already disposed are also in these seven. The ethyl group is known to be an entity and, on the generally (but not always) valid principle that fragmentation at an allylic bond is preferred, the $C_7H_{14}^{+}$ ion must include the double bond, whence $R = C_5H_{11}$ and $R' = H$, or $R = CH_3$ and $R' = C_4H_9$. The possible formulae are

$$\mathrm{CH_3{-}CH{=}CH\overset{+}{-}\underset{\overset{|}{C_4H_9}}{CH}{-}CH_2{-}CH_3},\ \text{or}$$

$$\mathrm{CH_3{-}CH_2{-}CH_2{-}CH_2{-}CH_2{-}CH_2{=}CH{-}CH_2{-}CH_2{-}CH_3}$$

The former should produce the C_7H_{14} ion by a re-arrangement following fission at a branch point. The requisite conditions are lacking in the other structure. The former, correct, arrangement must be preferred.

Again the argument seems to be extensive for the analysis of a rather simple structure. In many instances there is no alternative to a process of mutual exclusion and this will be the sole method available for small quantities (1–4 μg) of unstable or inaccessible materials. In other cases where the material is more abundant, readily available, and stable, it would be a fairly simple matter to reduce the bond and then analyse the alkane. A comparison of the two parent molecular ions would confirm that the original substance was an alkene. In the example, quantitative reduction is unnecessary provided that a pure specimen of the alkane is obtained.

The reduced product, an alkane, has a parent molecular ion corresponding to the formula $C_8H_{18}^{+}$. In accordance with arguments previously developed, it is preferable first to derive the skeletal structure. The alkane possesses an unusually abundant ion $m/e = 85$ corresponding to the elimination of an ethyl group at a branch in the chain; this is also true of the ion 71^+, which represents the loss of isopropyl. The ion 43^+ is the base peak of the spectrum, while $m/e = 41$ (26%) is much less abundant. Hence, in line with argument previously employed, the n-propyl group is considered to be less likely. The ethyl cannot be part of the isopropyl group unless there is an extensive re-arrangement, for which there is no evidence. Therefore, the two radicals are distinct and must, at best, be joined together. This accounts for at least five carbons; if the two groups are separately attached to a common carbon atom, then six. The presence of the parent molecular ion argues against a quaternary carbon, but the small abundance of P^+ favours a branched alkane. There is an abundant re-arrangement ion $C_5H_{10}^+$ which requires for its formation fission at a branch in the carbon chain plus an accompanying hydrogen migration. Because it is the only such ion of any abundance it is likely to arise from a fragmentation at the most crowded centre in the molecular structure, which is probably

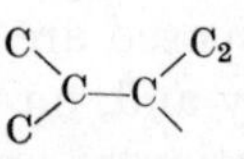

Two more carbons must be added, and the molecule is either

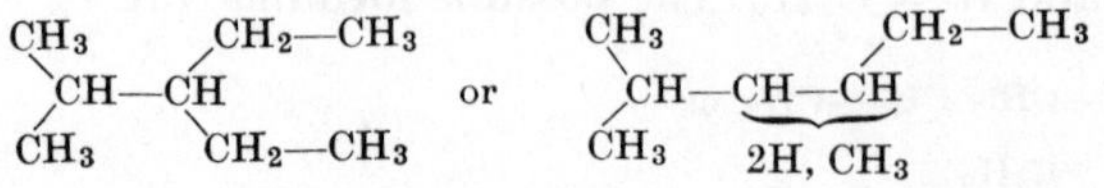

One other possibility exists. If the remaining unallocated carbon were placed adjacent to the isopropyl radical one might expect an abundant ion $m/e = 99$ resulting from the elimination of the single carbon branch. This is not the case. The alternative position should lead to the formation of an exceedingly abundant $C_4H_9^+$ ion as a consequence of the fission

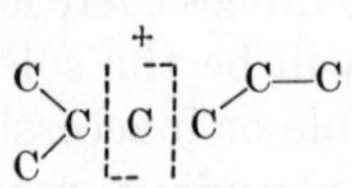

No very abundant ion $m/e = 57$ is observed.

It may be noted in passing that the isopropyl is a peculiarly stable grouping. If eliminated, as sometimes occurs readily, it may lose one or more hydrogen atoms. The loss of a methyl radical from it is rarely observed.

The bulk of the experimental evidence, therefore, indicates that the alkane is 2-methyl-3-ethylpentane. The analysis of the alkene from which it was derived now confirms the skeletal structure and positions the double bond. The unusually abundant ion $m/e = 83$ may be assigned to fission of the molecular ion at an allylic bond which must be terminal

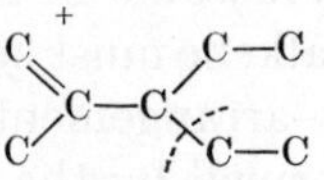

and which results in the facile elimination of an ethyl group. The alternative position which would still have the same effect is as shown

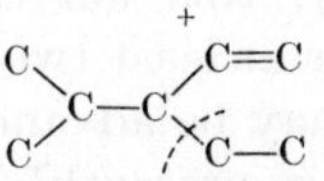

However, this second possible location would also favour the elision of the isopropyl group which, too, is allylically bonded. The opposite condition applies in the former since the isopropenyl group, being vinylically bonded to the remainder of the molecule, would not have

TABLE XVII

m/e	Abundance for ions of 2-methyl-3-ethylpent-1-ene (%)	Abundance for ions of 2-methyl-3-ethylpentane (%)	m/e	Abundance for ions of 2-methyl-3-ethylpent-1-ene (%)	Abundance for ions of 2-methyl-3-ethylpentane (%)
15	2·77	3·08	56	7·16	2·78
26	2·33		68	2·38	14·8
27	24·4	21·8	69	5·51	3·21
28	2·67	2·38	70	2·50	49·4
29	15·7	18·8	71	—	24·6
38	1·91		81	—	
39	27·3	12·0	82	2·07	—
40	4·60		83	56·0	—
41	46·0	26·4	84	29·9	4·19
42	5·64	14·0	85	—	17·9
43	10·8	100	86	—	—
44	—	3·33	97	2·72	—
51	3·12	—	112	6·90 (wavy underline)	—
53	8·15	2·24	113	0·59	—
54	—	—	114	0·02	1·25 (wavy underline)
55	100·0	17·5			

The parent molecular ion is marked thus 6·90 (wavy underline).

been lost easily. In fact the abundance of the ion $m/e = 41$ (46%) is not very great, but, more important, that at 43^+, and corresponding to the isopropyl group, is even less so (10·8%). It is certain that if the isopropyl group were allylically bonded it would be the base peak of the spectrum. Hence the formulation of the alkene must be 2-methyl-3-ethylpent-1-ene. The other rather abundant re-arrangement ion $m/e = 84$ is in agreement with the structure, as it is formed by the elimination of ethylene from one of the ethyl groups present.

The most remarkable feature of the observed alkene spectrum is the base peak $m/e = 55$ ($C_4H_7^+$); this ion cannot be obtained without molecular re-arrangement, or at least two bond fissions. Some more or less plausible explanations may be advanced, but convincing proof, or even satisfactory evidence is presently lacking. One possible interpretation has previously been made when considering the fission of the cycloalkanes, namely that alkenes and cycloalkanes may readily interconvert.

D. Bicycloalkanes

The bicycloalkanes represent one class of compound of the general formula C_nH_{2n-2}. As a class they are too varied for a concise analysis of group freatures, including as they do the various ring systems, and of course conformational variations.

Two general considerations seem valid, however, although they do not always advance the argument very far. It is reasonable to assume a preferred cleavage at a tertiary or quaternary centre. Which bond will break, if in fact only one does so exclusively, is a further problem. One may assume, on energetic grounds, that the bond, the rupture of which, will release most strain in the system and lead to the lowest free energy will be the one so favoured. Because many bicyclic structures are possessed of rigid conformations, the influence of *cis* and *trans* isomers, particularly in regard to the ring functions, will have a marked effect. Generally speaking, the more crowded structure will have the lowest stability, and consequently the least abundant parent molecular ion (Biemann and Seible, 1959).

Among the best known and most interesting pairs of isomers which have been examined are the *cis*- and *trans*-decalins. The *cis* isomer has the less abundant parent molecular ion both with respect to the base peak (when it is 71·8% against 85·5% for *trans*) and, more markedly, if P^+ is expressed as a fraction of the total ion current (Σ), when it is 0·067 as compared with 0·095. This is consistent with *cis*-decalin being the more strained, or congested, molecule. The number of 1,3-bond-interactions provides a rough and easily determined guide to strain in the molecule. For *cis*-decalin there are two, neglecting interactions within each six-membered ring, as these will be the same in each conformation. There are none in *trans*-decalin.

A further interesting point is that while the cracking patterns are very similar, the base peaks are different; for *cis*-decalin, the most abundant ion ($m/e = 67$) also represents 0·093 of the total ion current; for the other isomer, $m/e = 68$ (100%), although this is only slightly more abundant than $m/e = 67$ (96·3%). These two ions, expressed as a fraction of total ion current, become 0·111 and 0·107, respectively. It is clear, therefore, that in each isomer there is a facile reaction leading to the ion 67^+ and that in the *trans*-decalin there is a further even more facile decomposition

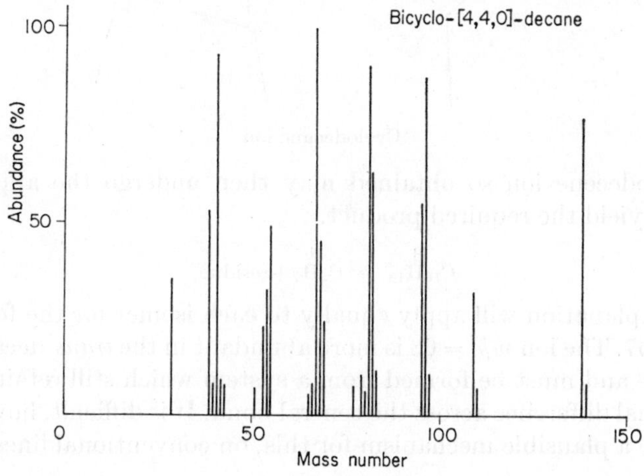

FIG. 15. Bicyclo-(4,4,0)-decane.

leading to 68^+, as the base peak. The common fragmentation process is assumed, in agreement with the arguments already advanced, to begin with cleavage at one of the ternary carbon atoms. The most favoured bond will be that connecting the two such centres present. Once this bond is broken, making the further reasonable assumption that one carbon will carry the charge, and as a carbonium ion can become planar, while the other (the radical) may readily invert configuration, the same ten-membered ring will be obtained from each isomer. Now it remains only to find a further cleavage to yield the desired product, and this may be achieved as follows. For a further fragmentation (that will yield the product), one may either reclose the system to yield a [5,3,0] ring, contrary to the principles already laid down or, by means of a *trans*- annular hydrogen shift, arrive at a cyclo-decene. This is likely since, in the cyclic decene system, carbons 1 and 5 approach each other closely.

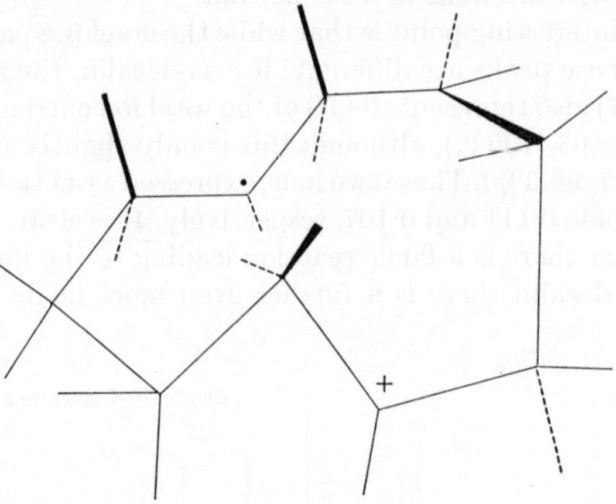

Cyclodecane ion

The cyclodecene ion so obtained may then undergo the appropriate fission to yield the required product.

$$\overset{+}{C_{10}H_{18}} \rightarrow \overset{+}{C_5H_7} + \text{residue}$$

This explanation will apply equally to each isomer for the formation of $m/e = 67$. The ion $m/e = 68$ is more abundant in the *trans*-decalin than in the *cis*- and must be formed from a system which still retains a conformational difference across the central bond. It is difficult, however, to formulate a plausible mechanism for this, on conventional lines.

E. Cycloalkenes and Alkenylcycloalkenes

There is little need for a detailed discussion of these systems. Cycloalkenes commonly give a very abundant parent molecular ion, which in itself is indicative of a cyclic structure; the molecular weight serves to confirm the general formula C_nH_{2n-2}. The presence of an alkyl chain is almost always revealed by the ready cleavage of the alkyl group leaving the ion corresponding to the ring very prominent, if not always the base peak. Even if the alkyl bond is vinylic to the double bond in the ring system, the alkyl chain is still readily cleaved from the cycloalkyl group. Fragmentation of substituted cyclopentene derivatives is as follows:

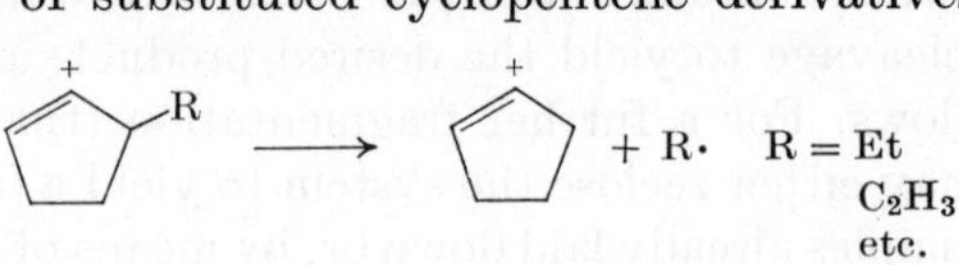

The presence of a double bond in the ring structure does not seem to affect the fission markedly, and, as has been noted previously, even when R = vinyl as in vinylcyclohexene, the base peak is still 81^+. It is interesting that, even in this case, there is a very abundant ion $m/e = 67$ which can be obtained, formally at least, by migration of the exocyclic double bond into the ring. Whether this is achieved by successive double-bond migrations, or by ring opening followed by ring-closure in a different manner, is uncertain. The important feature, and diagnostic tool, is that in many cases a compound which contains one ring and one double bond can often be recognized as such.

F. Dialkenes

The third group having the general formula C_nH_{2n-2} are represented by diolefins. The fragmentation pattern will obviously depend upon the relative distribution of the ethylenic groups. Thus molecular ions which are 1,5-dienes possess a doubly allylic bond which should break easily

$$\overset{+}{\text{R—C=C—C} \wr \text{C—C=C—R}'}$$

to yield one or more abundant ions depending upon the nature of R and R′. Molecules with ethylenic bonds more widely separated than this will yield fragments corresponding to the respective allylic fissions. The difficulties arise in the 1,4-diene in which an *a priori* estimate of the cracking pattern is difficult. The carbon–carbon bonds which lie between the double bonds are allylically disposed to one centre but vinylically to the other

$$\text{R—CH=CH—}\overset{+}{\text{CH}_2}\text{—CH=CH—R}'$$

and it is not easy to decide which factor outweighs the other. One point is, however, clear. The two hydrogens on the carbon atom between the double-bonds are doubly allylic and consequently a facile loss of hydrogen is to be expected. In addition to the parent molecular ion (P^+), there will be an abundant $(P-1)^+$, a rather uncommon occurrence. Experiments show that in addition to the usual fragment $C_3H_3R^+$ the ion $m/e = 42$ (R = H) is abundant. The ion must be formed by re-arrangement

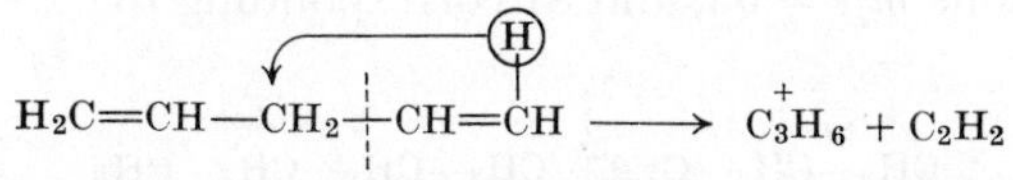

There is little doubt that the neutral product is acetylene. The energy required for such a process cannot be evaluated precisely since so many

terms in the equation are not accurately known. Substituting reasonable values shows, however, that the net energy change will be small.

Little information is available for the allenes, the 1,2-diolefins, but what there is suggests that the grouping $C_3H_3^+$ is rather stable and prominent in all the spectra. The only available monoalkyl allenes have rather abundant parent molecular ions and these appear to fragment in accordance with the following equations:

$$H_2C{:}\overset{+}{C}{:}CH.CH_2.R \longrightarrow \overset{+}{C_4H_5} + R\cdot$$

$$R.HC{:}\overset{+}{C}{:}CH.R' \longrightarrow R.\overset{+}{C_3H_2} + R'\cdot \quad \text{or} \quad R'\overset{+}{C_3H_2} + R\cdot$$

G. Acetylenes

The acetylenes constitute the last group of hydrocarbons of the general formula now under discussion. It is known from spectroscopic studies that for simple alkynes the bonds adjacent to the triple bond have a low

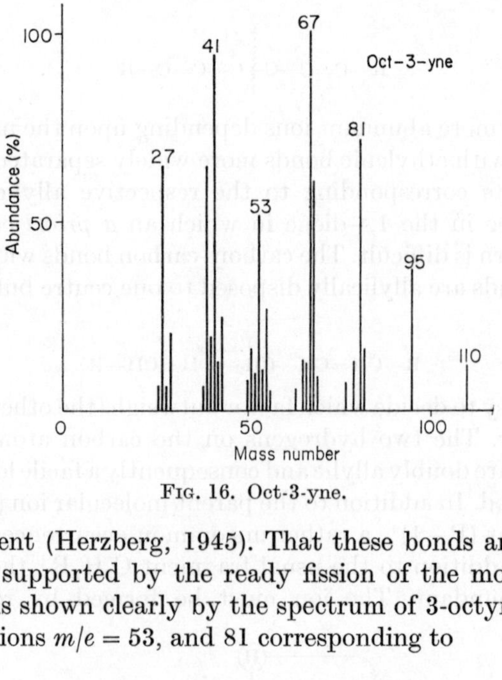

Fig. 16. Oct-3-yne.

bending moment (Herzberg, 1945). That these bonds are weak in the alkyne ion is supported by the ready fission of the molecule at such centres. This is shown clearly by the spectrum of 3-octyne where there are abundant ions m/e = 53, and 81 corresponding to

$$\overset{53}{CH_3{-}CH_2{-}}\,\overset{+}{C{\equiv}C}\,\big|\,\underset{81}{\overset{+}{CH_2}}{-}CH_2{-}CH_2{-}CH_3$$

as well as $m/e = 57$ and 95, the result of an allylic type fission namely

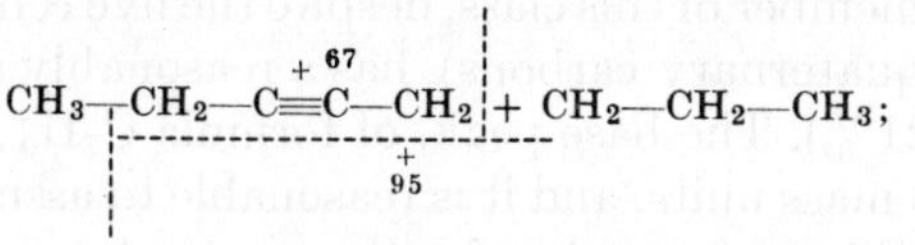

two of the abundant ions which remain, 39^+ and 41^+, may be obtained from the alkyl chain attached to the alkyne group. The ion $m/e = 27$ may arise from more than one centre and a detailed analysis is unjustified.

It may be of interest to point out that the ions $m/e = 53$, 67, 81, and 95 could arise by an allene type re-arrangement followed by a suitable fragmentation.

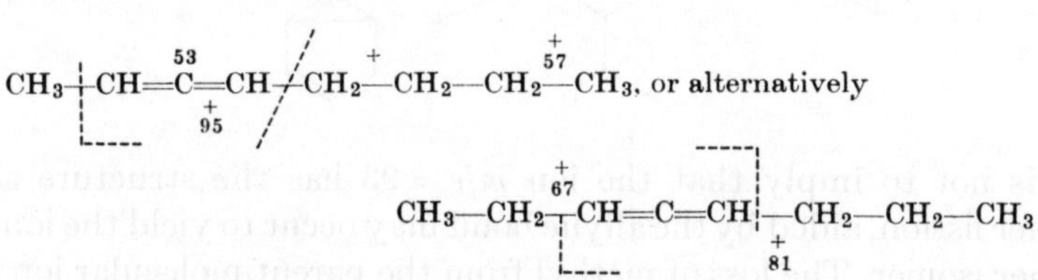

This can be investigated by the analysis of the mass spectra of other alkynes, and that of 4-octyne is suitable.

On the basis of direct fragmentations adjoining the triple bond one would expect two ions at $m/e = 43$ and 67. Assuming an intermediate formation of an allene type structure

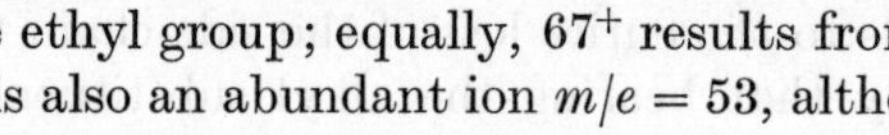

one would expect abundant ions at $m/e = 29$, 67 and 43, the last formed by the rupture shown with the charge occurring on the other fragment; this ion is absent from the spectrum. On the other hand, whilst there is no prominent ion at 29^+, there is the ion 81^+ corresponding to the loss of the ethyl group; equally, 67^+ results from the loss of the n-propyl. There is also an abundant ion $m/e = 53$, although its origin is not clear.

The corresponding fragments would, of course, be obtained if one assumed that both the bonds adjacent to the triple bond and those next removed were much weaker than average carbon–carbon bonds. Either interpretation assists in the analysis of the spectra of such compounds.

The general group of formula C_nH_{2n-6} is composed of many classes.

H. Tricycloalkanes

Tricyclene, a member of this class, despite the five centres of branching (including two quaternary carbons), has a reasonably abundant parent molecular ion (21%). The base peak, of formula $C_7H_9^+$, must be formed by the loss of 43 mass units, and it is reasonable to assume that they are lost by the breaking of two bonds. One may also postulate that the initial break occurs at one quaternary centre and that this is followed by break-down to a bicyclic structure and then the loss of the *trans*-annular *gem*-dimethyl group, present in the original molecule, together with a further hydrogen.

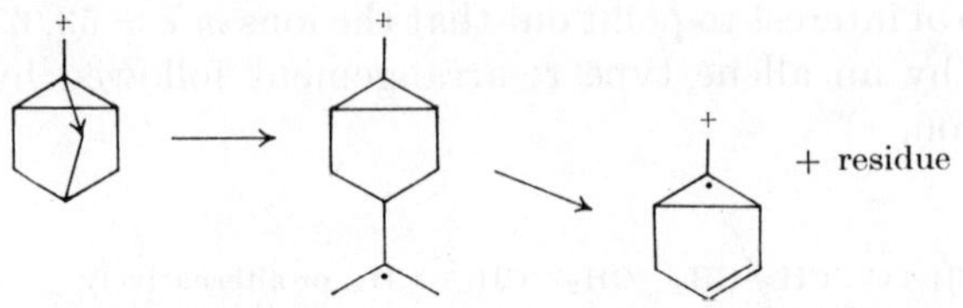

This is not to imply that the ion $m/e = 93$ has the structure shown. Further fission, aided by the allylic bond may occur to yield the ion $C_3H_4^+$ or other isomer. The loss of methyl from the parent molecular ion is also reported; but the site of fission, although clearly one of the quaternary carbon atoms, cannot be decided.

A survey of the terpenes suggests that the loss of the $(CH_3)_2C=$ group is fairly common, as also is that of a CH_3 group (McLafferty, 1963).

Another system which contains three fused rings is 5-n-pentadecyl-dodecahydroacenaphthene.

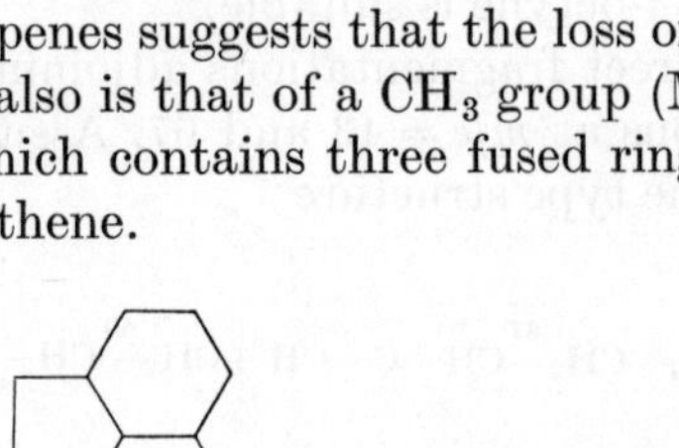

Again, the parent molecular ion is rather abundant, 18% of the base peak ($m/e = 121$; $C_9H_{13}^+$). Because the molecular weight is 374, the base peak corresponds to the entire loss of the side-chain together with one ring (presumably that bearing the alkyl substituent), and a further hydrogen. Two other ions $m/e = 81$ ($C_6H_9^+$; 62%), and 67 ($C_5H_7^+$; 49%) are also abundant.

The smaller ions are presumably derived from the base peak. Provided that two adjacent carbon atoms may pair the odd electrons to form a double bond (one electron has been removed to form the ion), further degradative processes which are consistent with known chemical

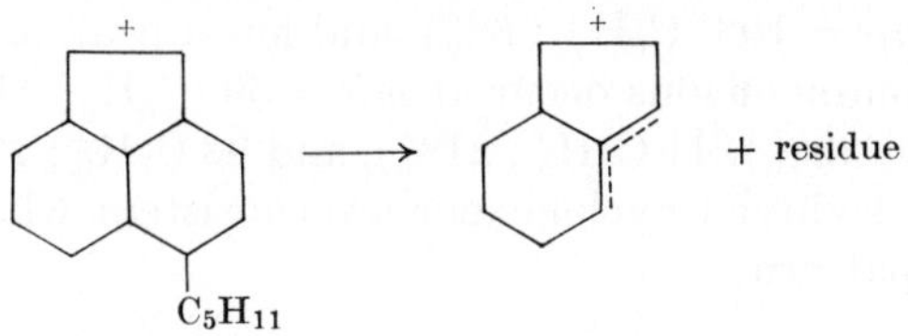

$m/e = 374$ $m/e = 121$

processes may be deduced. However, these may require further migration of the double bond and, while this is possible, there is no compelling evidence to require it; accordingly no detailed mechanism is proposed.

J. Tricyclic Alkanes

No mention has yet been made of systems which have one or more cycloalkyl groups attached to an alkyl chain of at least moderate length. The compound 1,7-dicyclopentyl-4-(2-cyclohexylethyl)heptane belongs to this class. The parent molecular ion is present, although weakly so, having an abundance of 0·6%. The base peak $m/e = 83$ corresponds to the cyclic group $C_6H_{11}^+$, consistent with the view that the loss of one cyclic system is a facile process.

K. Bicycloalkenes

Many natural products occur in this group and one, 2,6,6-trimethylbicyclo-[3,1,1]-heptene, better known as α-pinene, will be examined. The parent molecular ion 136^+ represents about one-twelfth of the base peak $m/e = 93$ ($C_7H_9^+$). The loss of 43 mass units is probably the result of the elimination of the C_3H_6 fragment, which is part of the four-membered ring, plus one other hydrogen. The two carbons are tertiary rather than quaternary but, on the other hand, one bond broken is allylically disposed. The loss of a methyl group is also observed in the ion $m/e = 121$ (13%), and this must be one of the *gem*-dimethyl groups. Other ions of moderate abundance are present at $m/e = 77$ (22·1%) and 79 (17·7%), the former $C_6H_5^+$, and the latter $C_6H_7^+$. Formulation of these ions is a difficult matter, and they probably arise from further degrading of the six-membered ring after the elimination of $\cdot C_3H_7$.

L. Alkenylcycloalkenes

As the number of double-bond equivalents increases, so also does the number of possible structures. The following example has a small molecular ion $m/e = 108$ ($C_8H_{12}^{+}$; 8%), and has a peak $m/e = 54$ ($C_4H_6^{+}$), whilst other prominent ions occur at $m/e = 39$ ($C_3H_3^{+}$; 41%), 56 ($C_5H_6^{+}$; 29%), 67 ($C_5H_7^{+}$; 23%), 41 $C_3H_5^{+}$; 21%), and 93 $C_7H_9^{+}$; 17%).

Such ions, in 4-vinyl-1-cyclohexene are consistent with the following fragmentation pattern:

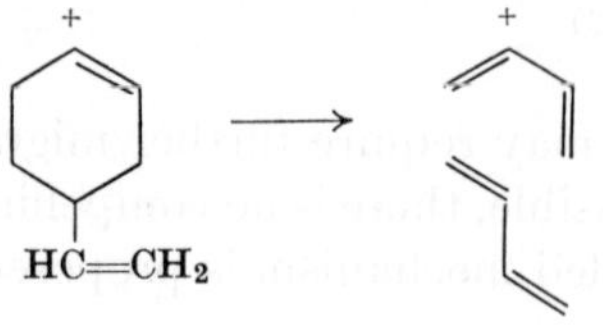

$m/e = 108$ $m/e = 54$

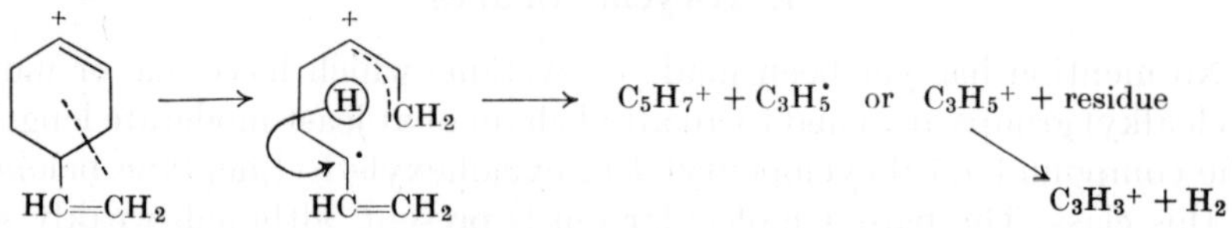

The first decomposition represents a retro-Diels-Alder, the second is a rupture at a doubly allylic bond—a particularly favoured fragmentation.

M. Alkenynes

Again one example must suffice. The parent molecular ion $m/e = 66$ is also the base peak of the spectrum of *trans*-pent-2-en-4-yne. The high stability is related to the structure; the carbon bond joining the acetylene residue to the remainder of the molecule is vinylic. The characteristic feature of acetylenes already noted, namely the loss of one hydrogen, occurs here also. The next most abundant ion 39^+ ($C_3H_3^{+}$) is undoubtedly formed by the fission of the propenyl group, which loses a further two hydrogens to give what is either the cyclopropenyl ion or $CH_2{-}C{\equiv}CH^+$.

$$\overset{+}{\mathrm{CH_3{-}CH{=}CH}}\,\vdots\,\mathrm{C{\equiv}CH} \longrightarrow \overset{+}{\mathrm{C_3H_3}} + \mathrm{C_2H} + \mathrm{H_2}$$

Neglecting the remaining isomers of this and other classes which do not contribute anything new to the established modes of fragmentation, only one other acyclic system need be discussed, viz. the alkadiynes of formula C_nH_{2n-6}. The parent molecular ion is usually abundant, if not the base peak of the spectrum for the simpler members, but its abundance falls off as the aliphatic chain increases in length. The loss of a single hydrogen is also likely as the $(P-1)^+$ ion is some 20–30% of the base peak. In the case of 1,5-hexadiyne this ion $m/e = 39$ corresponds to the reaction

$$HC{\equiv}C{-}\overset{+}{C}H_2{-}CH_2{-}C{\equiv}CH \rightarrow \overset{+}{C_3}H_3 + C_3H_3$$

Such a fragmentation cannot occur in 2,4-hexadiyne where the acetylene groups are adjacent: then the molecular ion yields a fragment $m/e = 52$ ($C_4^+H_4$), which is almost equally abundant in the 1,5-isomer already mentioned (57%). The ions $m/e = 51$ and 50 are comparable in abundance with 52^+. From this, one deduces that the $C_4H_4^+$ ion most probably contains an acetylene unit; even so, the structure of the ion is uncertain. In the case of 1,5-hexadiyne, the derivation is not difficult, but considerable re-arrangements would be required in other isomers.

N. Aromatic Compounds

These form an extensive and exceedingly important group of hydrocarbons. The aromatic compounds: benzene, naphthalene etc., show considerable stability; a stability which increases as one moves higher in the series. Benzene itself yields ions corresponding to $(P-1)^+$, $(P-2)^+$, $(P-26)^+$, $(P-27)^+$ etc., and some of these are of considerable abundance. Naphthalene is rather more stable, but when one examines chrysene, benzanthracene, and the polynuclear aromatics in general, the fragment ions become even less abundant. The parent molecular ion continues to be the base peak of the spectrum, and now the second most abundant is the doubly ionized parent molecular ion. The loss of a single hydrogen is common leaving a $(P-1)^+$ ion more than 10% the base peak. The $(P-2)^+$ ion is often still more prominent and there is also an ion $(P-26)^+$ (5%); the latter ion corresponds to the loss of acetylene. It is sometimes useful in detecting the presence of polynuclear aromatics in more elaborate structures (Reed, 1960).

In addition to polynuclear compounds of the general series anthracene, tetracene, and their variants phenanthrene, benzanthracene, perylene, pyrene etc., there is a further series of polyphenyls. Very little study has been made of these but they too have a parent molecular ion which is the base peak; they also exhibit $(P/2)^+$, $(P-1)^+$ and $(P-2)^+$ ions. The present, slender evidence suggests that, in this series, the abundance of the $(P-1)^+$ ion is greater than that of $(P-2)^+$.

P. Alkylated Aromatic Hydrocarbons

Substitution on the ring of one or more alkyl groups has a significant effect upon the cracking pattern. The base peak of many spectra, viz. toluene, ethylbenzene, propylbenzene, isomeric xylenes, and similar structures, is now $m/e = 91$. Theoretical reasons have been adduced which substantiate the view that the ion formed is the tropylium ion (Meyerson and Rylander, 1957, 1958). When more than one substituent is present, as with *p*-xylene, the molecular ion loses methyl, or at least its elements, to yield the tropylium ion. This tendency to yield a common abundant ion by fission at the benzylic bond is not so marked in naphthalene derivatives. The parent molecular ion of 2-methylnaphthalene is also the base peak; that corresponding to the loss of hydrogen $(P-1)^+$, although abundant (67·1%), is less important in the spectrum. Moreover, the ion corresponding to the elimination of methyl $(P-15)^+$ has become prominent (23%). In the case of polysubstituted naphthalenes, although ions formed by fission of the bond β to the aromatic nucleus are prominent, there is little evidence to support any suggestion that the formation of a benzotropylium ion plays an important part in the fragmentation process.

Alkyl substituents also influence the cracking patterns of the polyphenyls. As may be expected, loss of substituents by cleavage at the β-bond again occurs. In 3-methylbiphenyl, $(P-1)^+$ is some 40% of the base peak which is also the parent molecular ion. Again the ion $(P-15)^+$ is reasonably prominent and $(P-16)^+$ is more abundant still (17%); it may well be, therefore, that the formation of the tropylium ion is not the only driving force in the elimination of substituent groups from off the benzene nucleus. Another point of interest is the rather ready loss of a further two hydrogen atoms to yield $(P-3)^+$ (20%). This, apart from considerations of the molecular formula, is sufficient to categorize a substituted biphenyl.

Q. Hydroaromatic Compounds

In view of what has been said previously, it might be inferred that, in the tetralins at least, the ion 91^+ would be obtained with the complete elimination of all but one carbon of the allylic system. This does not seem to be true, although there is present the ion $m/e = 91$ in moderate abundance in many cases. In 1-ethyl-(1,2,3,4)-tetrahydronaphthalene it amounts to some 15% of the base peak. Here also the base peak corresponds to the loss of twenty-nine mass units. Unfortunately, the origin of this fragment has not been proved, but surely is as follows:

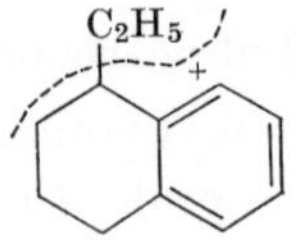

$\longrightarrow \overset{+}{C_{10}H_{11}} + \cdot C_2H_5$

$m/e = 160$ $\qquad$ $m/e = 131$

Both bonds that are broken are allylic.

In view of this fragmentation, that of 2a,3,4,5-tetrahydroacenaphthene seems relatively straightforward. The parent ion $m/e = 158$ ($C_{12}H_{14}^{+}$; 35%) is fairly prominent; the base peak $m/e = 130$ ($C_{10}H_{10}^{+}$) results from the loss of ethylene, which may well be from a retro-Diels Alder reaction.

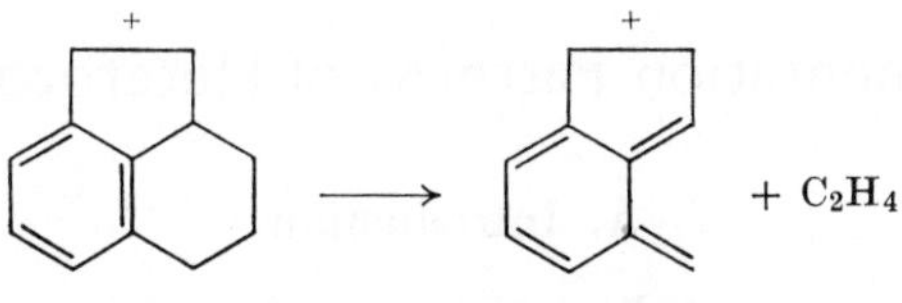

$m/e = 158$ $m/e = 130$

Again, it is not an easy matter to determine the origin of the other moderately abundant ion $m/e = 115$ ($C_9H_7^{+}$; 23%). It represents the loss of forty-three mass units from the parent molecular ion, or, alternatively, a further methyl group from the base peak. A mechanism which requires double-bond migration may be devised for the loss of the methyl group

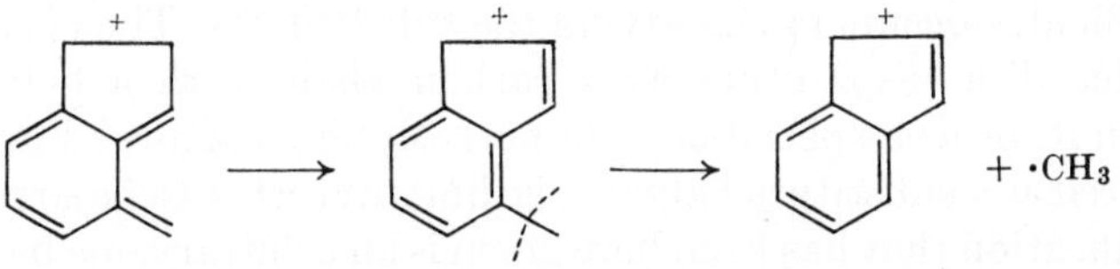

but this does not seem very likely, in view of the strong bond which connects the methyl to the aromatic centre. Similar difficult problems occur when a rational explanation of the loss of a $\cdot C_3H_7$ group is sought.

The observation that the ions $(P-29)^{+}$ (30%) and $(P-30)^{+}$ (15%) are fairly prominent again raises problems of interpretation, particularly because there is (in the absence of experiments upon suitably labelled compounds) no means of knowing which atoms are removed. Certainly, if the ion $m/e = 128$ represents a naphthalene ion, very many re-arrangements would be required to obtain it; complicated migrations have, however, often been demonstrated in aromatic compounds (Beynon, 1960, p. 70). Great interest would attach to studies in the cracking patterns of carbon-13 labelled aromatic substances.

CHAPTER 4

The Fragmentation Patterns of Heterocompounds

A. Introduction

The complete understanding of the cracking patterns of the hydrocarbons, while of immense importance, would only provide a part solution to the determination of organic structures. There remains the equally complex problem of identifying both the nature of any substituent groups and what is often very much more difficult, the position of these upon the carbon skeleton. Some suggestions as to the methods to be adopted will be given later (see Chapter 7, p. 143). Before these can be applied some attention must be given to the effect of substituents upon the cracking pattern of any hydrocarbon system.

One difficulty occurs in classifying the substituents. Thus is it correct to consider di-n-hexyl ether as a carbon chain with a hetero atom inserted in it, or as a special class of heterolytic structure? Further if it is considered as a substituted alkane, is dimethyl ether to be so regarded? The classification that has been here given is an arbitrary one based upon expediency. The cracking patterns of acyclic ethers depend to some extent upon the nature of the carbon skeleton surrounding the oxygen atom. The most distinctive and atypical behaviour stems from a one-carbon chain—a methyl group. Moreover, the methoxyl group is common in natural products; it is best to treat methoxyl ($CH_3O\cdot$) as an entity. The problem of ethoxyl is more difficult; it is by no means as common as butoxyl or higher alkoxy groups, nor is the fragmentation pattern of ethoxyl so very different, after making allowance for the different number of carbon atoms. Ethoxyl and higher homologues are, therefore, considered as systems in which the oxygen atom has been introduced into the carbon chain, while methoxyl is considered an entity. Such arbitrary decisions have been kept to a minimum, being restricted to amines and the simplest ether group. Other systems which are treated as single entities include: amide, cyano, nitro, nitroso, isonitroso, the carboxylic group and simple esters. More elaborate systems are treated as hetero-atoms introduced into an otherwise hydrocarbon structure.

As is well known from general chemical experience, the presence of groups of the kind just discussed has a marked effect upon electron distribution, even in a neutral molecule. Groups such as methyl increase the electron density upon a neighbouring group, as is shown by an arrow directed away from the methyl towards the particular group $CH_3 \rightarrow CH_2-$; groups such as the hydroxyl act in the opposite sense

$$HO \leftarrow \underset{\displaystyle CH_3}{\underset{|}{CH}} - CO_2H$$

These considerations are complicated by the presence of other effects in conjugated and, especially, aromatic systems. The further problems of electron distribution are presently ignored, at least in the discussion upon substituted alkanes. Because the lowering of the electron density in a bond will facilitate its fragmentation, the electron distribution will obviously affect the stability of the parent molecular ion. This consideration is reinforced by the fact that, since mass spectrometry is mainly a study of positive ions (at its present state of development), the species under examination will also be electron deficient from their method of production. The influence of the electron displacement, the inductive effect, has already been extensively examined for neutral molecules, and it has been determined that its magnitude diminishes rather rapidly along a saturated carbon chain. The effect is most marked for the α-bond in the following structure, less for β, still less for γ; it is sensibly the same in all the others.

$$\underset{\alpha}{Cl \leftarrow} \underset{\beta}{CH_2 \leftarrow} \underset{\gamma}{CH_2 \leftarrow} \underset{\delta}{CH_2 -\!\!-} CH_2 \text{ etc.}$$

Now, if the same property exists in the positive ion and is not masked by the electron deficiency which results from ionization, one could predict the consequence of introducing a substituent into an alkane in its effect upon the mass spectrum (McLafferty, 1955). This possibility has been investigated by comparing the spectrum of n-hexane with those of pentan-1-ol and pentane-1-thiol. Regarding n-pentane as the reference compound; this has been successively substituted by a methyl group, which may increase the electron density in the $C_5H_{11}\cdot$ radical, a hydroxyl group and a thiol group, which have an inductive effect in the opposite sense.

For comparison purposes, the more abundant ions which are included in Table XVIII have been expressed as a proportion of the total ion current rather than the base peak. It will be observed that the values of this total are nearly the same in all three cases, and the qualitative conclusions are unaffected if the abundances of the ions are expressed as a percentage of the base peak instead.

The mass spectra may be considered in two parts. There are the fragmentations which occur in the molecule and are remote from the substituent. These, which will include the fragment ions $m/e = 27$, 29, 39, 40, and 41, may be expected to be the same if the inductive effect of the substituent does not extend more than three carbons. The figures

TABLE XVIII

Σ	521·7	545·8	502·9	391·6
26				1·39
27	8·70	2·00	6·82	8·25
28		2·69	1·26	1·25
29	11·62	12·16	6·74	6·89
30		0·47		
31		12·62		
32				
35			0·86	
39	3·73	3·34	3·72	8·81
40	0·61	0·63	0·65	1·19
41	13·43	11·05	8·17	11·49
42	7·84	18·32	19·89	25·54
43	13·56	5·22	4·93	1·06
45		1·01	2·11	
47			2·53	
				1·19
54				
55	1·25	10·88	9·3	14·79
56	8·68	2·55		0·66
57	19·18	3·94	0·75	
61			2·72	
70		6·72	7·30	8·10
71	0·95	0·84	0·67	
86	2·97		6·44	
m/e	n-Hexane	Pentan-1-ol	Pentane-1-thiol	Pent-1-ene

recorded in Table XVIII support this conclusion although the agreement in some instances is more of an order of magnitude than a precise one.

At the other end of the scale, consideration must be given to the fragments that arise by cleavage of the carbon–carbon bonds in the neighbourhood of the substituent. Here it is more difficult to obtain a reasonable estimate of the inductive effect because, for the fragmentation

$$\overset{+}{R}CH_2X \longrightarrow \overbrace{R + CH_2X}^{+}$$

to take a particular example, the position of the charge in the product will depend upon the relative ionization potentials of R• and •CH_2X. It is not possible, in the present state of our knowledge, to distribute the positive charge fairly between them. Therefore, in order to have a measure which will possess some qualitative significance, the ion abundances of the two fragments have been summed. This is an empirical solution to the problem which is easy to carry out; for the fission process shown in the equation one must add the ion abundances of CH_2X^+ and P—CH_2X^+, $CH_2CH_2X^+$ and P—$CH_2CH_2X^+$, etc. These summations yield the results shown in Table XIX.

TABLE XIX

Compound	n-Hexane	Pentan-1-ol	Pentane-1-thiol
$\text{P—}\overset{+}{\text{C}}\text{H}_2\text{X} + \overset{+}{\text{C}}\text{H}_2\text{X}$	2·0	13·46	6·28
$\overset{+}{\text{P}}\text{—C}_2\text{H}_4\text{X} + \text{C}_2\overset{+}{\text{H}}_4\text{X}$	30·80	6·23	7·65
$\overset{+}{\text{P}}\text{—C}_3\text{H}_6\text{X} + \text{C}_3\text{H}_6\text{X}$	13·56	2·00	6·82

It is evident that there is no simple correlation between the inductive effect of individual groups and the mass spectrum. Several reasons may be proposed for this, some of which are mentioned below; however, since they do not advance the identification of organic compounds by mass-spectrometric methods, they will not receive a detailed discussion.

There is evidence that some of the hydrocarbon ions present derive from re-arrangements that occur when once the fragment is cleaved off; the ion $m/e = 43$ is considered to be a protonated cyclopropane ion, and this will have a different enthalpy from the n-propyl ion. It may be that re-arrangements are possible in the corresponding hetero-ions, e.g. $CH_2CH_2OH^+$ could re-arrange to CH_3CHOH^+, and the simple analysis proposed would be too poor a representation of the actual state of the system to allow predictions to be made.

The suggestion has been advanced in the restricted field of hydrocarbons that the removal of an electron upon ionization allows considerable delocalization of the remaining electrons, even with an alkane ion (Lennard-Jones and Hall, 1952). The theory has been successful in determining the energy levels in n-octane. Should this delocalization occur, effects other than the inductive effect may operate. Since one centre in the ion must be electron deficient, charge distributions may include the unshared pair electrons on the oxygen, and it will be difficult to allow for this.

If the above postulate is even partly correct, then the replacement of the oxygen atom by sulphur will alter the condition of the ion in one important respect. Sulphur and other atoms of the same and higher periods possess *d*-shell electrons; these may also play a part in the delocalized orbitals of the molecular ion, and thus intervene decisively in the fragmentation processes.

A further possibility exists, namely that, either as a consequence of the thermal conditions of the sample or as a result of electron impact, the molecular ion formed is substantially if not exclusively that of pent-1-ene. Therefore, comparisons between the molecules containing hetero-atoms which can eliminate water or hydrogen sulphide, and n-hexane which cannot are invalid. The spectrum of pent-1-ene has been included in Table XVIII for the purpose of comparison. It will be evident that, while there are some resemblances in the spectra, they do not sustain a detailed comparison.

The above discussion overlooks another common phenomenon which causes further problems in analysis; compounds such as are discussed in this book often participate in another type of re-arrangement which is concomitant with or immediately sequential to the fission of the β-bond. One example, that of capronitrile, has already been given; a further one is the fragmentation, with hydrogen migration, of the valeramide ion.

Fragmentations

1. ALIPHATIC COMPOUNDS

A. α-Bond fission

$$\mathrm{R\underset{\cdot}{C}H_2\overset{+}{C}H_2X \rightarrow RCH_2\overset{+}{C}H_2 + X\cdot}$$

$\mathrm{X = OR', SR', SH, NRR', NC, Br, and\ I}$

B. β-Bond fission

Without re-arrangement

$$\mathrm{R\overset{+}{C}H_2 \,(\, CH_2Y \longrightarrow R\overset{+}{C}H_2 + \cdot CH_2Y}$$

$\mathrm{Y = OH\ and\ NCS}$

With re-arrangement

$$\mathrm{R\underset{\underset{H}{|}}{C}HCH_2 \mid CH_2Z \longrightarrow R\overset{+}{C}H{=}CH_2 + \cdot CH_3Z}$$

$\mathrm{Z = CHO, COR, -SH, -NH_2, -CONRR', -CN, F, and\ Cl}$

2. AROMATIC COMPOUNDS

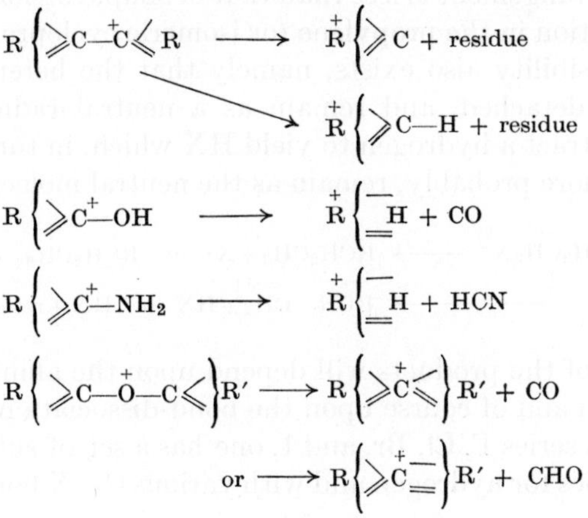

$$R\left\{\begin{array}{c}\\\\\end{array}\right. \overset{+}{C}-\overset{+}{C}\left.\begin{array}{c}\\\\\end{array}\right\langle R \longrightarrow \overset{+}{R}\left\{\begin{array}{c}\\\\\end{array}\right. \overset{+}{C} + \text{residue}$$

$$\longrightarrow \overset{+}{R}\left\{\begin{array}{c}\\\\\end{array}\right. \overset{+}{C}-H + \text{residue}$$

$$R\left\{\begin{array}{c}\\\\\end{array}\right. \overset{+}{C}-OH \longrightarrow \overset{+}{R}\left\{\underline{}\atop H\right. + CO$$

$$R\left\{\begin{array}{c}\\\\\end{array}\right. \overset{+}{C}-NH_2 \longrightarrow \overset{+}{R}\left\{\underline{}\atop H\right. + HCN$$

$$R\left\{\begin{array}{c}\\\\\end{array}\right. C-O-C\left.\begin{array}{c}\\\\\end{array}\right\langle R' \longrightarrow R\left\{\begin{array}{c}\\\\\end{array}\right. \overset{+}{C}\left.\begin{array}{c}\\\\\end{array}\right\rangle R' + CO$$

$$\text{or} \longrightarrow R\left\{\begin{array}{c}\\\\\end{array}\right. \overset{+}{C}{=}\left.\begin{array}{c}\\\\\end{array}\right\rbrace R' + CHO$$

$$\overset{+}{R}CHO \rightarrow H + \overset{+}{R}CO$$
$$\rightarrow \overset{+}{R} + CO$$

$$\overset{+}{R}COR' \rightarrow \overset{+}{R} + \text{residue}$$
$$\rightarrow \overset{+}{R'} + \text{residue}$$
$$\rightarrow \overset{+}{R}CO + \text{residue}$$
$$\rightarrow \overset{+}{R'}CO + \text{residue}$$

$$\overset{+}{R} CO OH \begin{array}{c} \nearrow \overset{+}{R} + \overset{\cdot}{H}CO_2 \\ \searrow \overset{+}{R}CO + \cdot OH \end{array}$$

$$CH_3\overset{+}{C}HCH_2{-}CH_2CONH_2 \longrightarrow CH_3CH{=}\overset{+}{C}H_2 + \text{residue}$$
$$\text{or isomer}$$

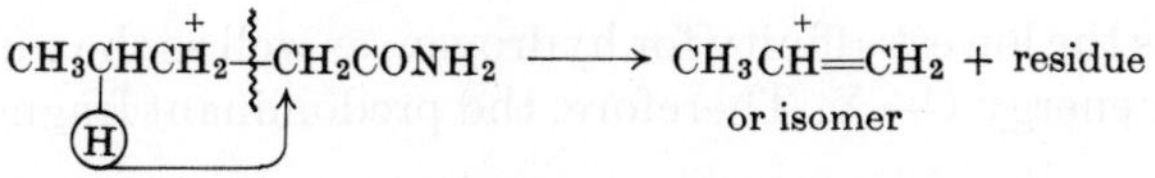

Now re-arrangements of this type are energetically more favourable than simple fissions for, in addition to the fragmentation which is common to both, they break a further bond, but make two. If we assume (and this is a little unfavourable to the argument) that the bond-dissociation energy needed to remove the hydrogen from the carbon chain is the same as that released on forming acetamide, then the energy of formation of the $C_3H_6^+$ molecular ion can be set against that of the original fission. Hence the energy demand of fragmentation with concomitant re-arrangement is less than that of simple fission by the energy of bond formation in the propylene (or isomeric cyclopropane) ion.

A third possibility also exists, namely that the heterogroup X will itself become detached, and remain as a neutral radical, a positive ion X^+, or abstract a hydrogen to yield HX which, in turn, may appear as an ion or, more probably, remain as the neutral molecule.

$$RCH_2\overset{+}{C}H_2X \longrightarrow R\overset{+}{C}H_2CH_2 + X\cdot \quad \text{or} \quad RCH_2CH_2\cdot + \overset{+}{X}$$

$$\longrightarrow R\overset{+}{C}H{=}CH_2 + HX \quad \text{or} \quad \overset{+}{H}X + \text{residue}$$

The nature of the products will depend upon the affinity of X· for a hydrogen atom and of course upon the bond-dissociation energy C—X. In the halogen series F, Cl, Br, and I, one has a set of substituents with varying affinities for hydrogen and with various C—X bond-dissociation energies.

TABLE XX

X	D(H—X)	$D(CH_3$—X)	$D(C_2H_5$—X)	D(n-C_3H_7—X)	D(i-C_3H_7—X)
H	4·58	4·44	4·09	4·34	4·08
F	6·47	—	—	—	—
Cl	4·50	3·35	3·13	—	—
Br	3·82	2·92	2·68	—	—
I	3·13	2·30	—	—	—
CN	4·94	3·31	—	—	—
OH	5·09	4·53	3·72	—	—
OCH_3	4·35	3·82	—	—	—

These values were obtained either directly or by standard methods from the information in Cottrell, T.L., 1958 and A.P.I. Project-Chemical Thermodynamic Properties.

Iodine has the lowest affinity for hydrogen, as well as the weakest bond dissociation energy C—X. Therefore, the predominant fragmentation is

$$\overset{+}{R}X \rightarrow \overset{+}{R} + X$$

At the other extreme is fluorine, where the affinity for a hydrogen is very great and in which elimination occurs.

$$RC\overset{+}{H}_2CH_2F \rightarrow RC\overset{+}{H}{=}CH_2 + HF$$

For chlorides this process is if anything more marked. The C—Cl bond is considerably weaker than in the fluoride but the H—Cl bond is of the same order of magnitude as in HF which means that elimination of hydrogen chloride will be favoured relative to the loss of chlorine.

When bromine is the substituent, the behaviour corresponds more closely to that of iodine, as shown in Table XXI.

Other groups in which this further effect is possible are hydroxyl and, in favourable circumstances, methoxyl.

The nitriles are known to eliminate hydrocyanic acid and, occasionally, a similar type of re-arrangement has been noted for methyl esters. These reactions will be referred to again in more detail in a later section of this chapter (p.83).

There is a class of compound in which the inductive effect is of the opposite sign. Amines are typical of this group and, in the neutral molecule, they strengthen the neighbouring carbon–carbon bonds by electron accession. On ionization, however, when an electron is removed, probably from the nitrogen, the formation of the positive ion will weaken the bonds; the two forces are opposed. The general outcome is that, for the ions of long-chain aliphatic amines, this substituent behaves, by and large, as an electronegative group,

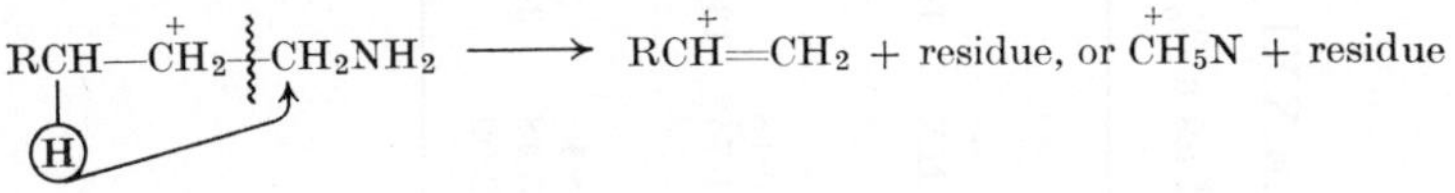

with β-fission and a concomitant hydrogen migration. The electronegative property may be marginal, however, for in secondary or tertiary amines the fragmentation is often at the α-bond with the loss of the longest branch,

$$R_2\overset{+}{R'N} \rightarrow R_2\overset{+}{N}H + \text{residue} \qquad R' > R$$

analogous to fission at a point of branching in an alkane.

A further mechanism has also been proposed (Collin, 1952, 1954); the amine is considered to protonate, giving the ammonium ion $R_2R'NH^+$. This is likely, since amines are basic and the mass spectrometer may be considered in a peculiar sense as an acid medium. The "onium" ion so formed will then, by formal analogy to conventional chemistry, undergo a Hofmann (1851 a,b) elimination, yielding an alkene. This is consistent with fission at the α-bond which, as already reported, is a common feature of the fragmentation of some amines. It is, of course, compatible with the removal of the longest alkyl substituent as an alkene containing all the carbon atoms originally present. The Hofmann elimination is known to favour the removal of the least alkylated carbon chain. A suitably constructed molecule would provide a test of the correctness of this

TABLE XXI

Abundances expressed as a % of the base peak

X	M = CH_2				M = C_2H_4				M = n-C_3H_6				M = i-C_3H_6			
	$\overset{+}{\mathrm{HM}}$	$\overset{+}{\mathrm{X}}$	$\overset{+}{\mathrm{M}}$	$\overset{+}{\mathrm{HX}}$	$\overset{+}{\mathrm{HM}}$	$\overset{+}{\mathrm{X}}$	$\overset{+}{\mathrm{M}}$	$\overset{+}{\mathrm{HX}}$	$\overset{+}{\mathrm{HM}}$	$\overset{+}{\mathrm{X}}$	$\overset{+}{\mathrm{M}}$	$\overset{+}{\mathrm{HX}}$	$\overset{+}{\mathrm{HM}}$	$\overset{+}{\mathrm{X}}$	$\overset{+}{\mathrm{M}}$	$\overset{+}{\mathrm{HX}}$
H	85·6	—	17·1	—	21·2	—	100	—					(23·1)**	—	(5·97)**	—
F	100	2·04	17·2	0·34	5·69	0·24	6·04	0·18					1·22	—	1·58	—
Cl	83	5·56*	7·78	1·46*	83·60	2·96*	90·9	1·74*	13·8	0·44*	100	1,75*	100	0·46*	7·28	1·70*
Br													100	2·35†	5·22	0·84†
I	12·8	37·8	0·92	2·70												
CN					2·70	20·5	100	18·4								
OH					22·48	—	5·49	5·48	3·05		7·17	0·48	16·64	—	3·96	0·48
OCH_3	37·36	7·04	12·49	0·13	49·04	19·85	7·88	2·09					29·83	18·47	9·32	0·55

* $^{35}\overset{+}{\mathrm{Cl}}$ and $\mathrm{H}^{35}\overset{+}{\mathrm{Cl}}$.

† $^{79}\overset{+}{\mathrm{Br}}$ and $\mathrm{H}^{79}\overset{+}{\mathrm{Br}}$.

** Structure and origin of these ions uncertain.

Values are taken from the A.P.I. Project 44.

method of forming the ion, but so far adequate information is not available. The main drawback to the idea of protonation would seem to be that mechanisms of this type would be bimolecular and would not have a linear dependence upon the source pressure of the sample.

B. Aliphatic Halides

The halides, as already discussed, seem to lead to α-fission with or without hydrogen abstraction; β-fissions do occur. The most extensive series studied is that of the perfluorohydrocarbons (Mohler, Bloom, Lengel and Wise, 1949). Allowing for molecular weight differences, the spectrum obtained resembles that of the corresponding alkane. Perhaps the most distinctive difference is the ready elimination and rather high abundance of the trifluoromethyl ion ($m/e = 69$).

C. Alcohols

The alcohols, one of the groups of oxygenated compounds reported by Thomas and Seyfried (1949), show much similarity in cracking patterns with those of the alkenes, which is consistent with the view that the elimination of water occurs easily. This may be as a result of the thermal cracking of the neutral molecule to the olefine which is then ionized, or, alternatively, elimination from the parent molecular ion may occur.

$$HOC\overset{+}{H_2}CH_2R \rightarrow CH_2{=}\overset{+}{C}HR + H_2O$$
or isomer

In addition to olefine formation, the primary alcohols show the ready loss of thirty-three mass units, presumably CH_3OH and H_2 but possibly some other combination. In the spectrum, the parent molecular ion P^+ is often only weakly abundant and moreover, becomes even less prominent as the molecular weight increases.

Tertiary alcohols do not eliminate water on ionization but, in common with tertiary amines and branched alkanes, eliminate the largest alkyl radical

$$R''R'\overset{+}{R}COH \rightarrow R'R\overset{+}{C}OH + \cdot R'' \qquad R'' > R', R.$$

The problem of placing a hydroxyl group along the carbon skeleton, an important analytical feature, is deferred to a later chapter (see Chapter 8, p. 178).

D. Trimethylsilyl Ethers

Alcohols may readily be converted into the corresponding trimethylsilyl ethers which are more stable than the original alcohol and very much more volatile. Elimination resulting from thermal cracking is no longer a problem. The silyl ethers, in common with alkanes, possess a quaternary carbon atom and do not have a parent molecular ion or at least only a very feeble one. However, the base peak corresponds to the loss of a methyl radical from the trimethylsilyl group; because of this, they are useful for determining the molecular weight of the original alcohol.

$$\mathrm{RO\overset{+}{S}i(CH_3)_3 \rightarrow RO\overset{+}{S}i(CH_3)_2 + \cdot CH_3}$$

The general pattern of the alkyl chain R can usually be obtained by application of principles already outlined. Thus, if R is a secondary group of the form

$$\mathrm{\begin{matrix} R' \searrow \\ \quad \overset{+}{C}HO \\ R'' \nearrow \end{matrix}}$$

where R″ > R′, then preferred elimination of R″ occurs.

The general similarity of cracking patterns amongst the silyl ethers, which is particularly marked for secondary and tertiary compounds, makes it difficult to distinguish between these (Sharkey, Jr., Friedel and Langer, 1957).

E. Methoxyl

The second method of preventing the elimination of water from an alcohol is to convert it to a methyl ether. The observed method of fragmentation then becomes

$$\mathrm{CH_3\overset{+}{O} \wr CH_2R \longrightarrow CH_3\overset{+}{O} + residue}$$

Fission at a carbon–oxygen bond with the elimination of a hydrogen and probable formation of methanol (or hydrogen and formaldehyde) also occurs, provided that there is no branching on this carbon.

$$\mathrm{CH_3O \wr CH_2\overset{+}{C}HR \longrightarrow R\overset{+}{C}H{=}CH_2 + CH_3OH}$$

Ethers seem rather sensitive to the nature of the adjacent carbon skeleton. Consequently, they are useful for the diagnosis of structure. The parent molecular ion is more abundant than in either the alcohol or the trimethylsilyl ether of comparable structure, which makes it useful in the determination of molecular weight (McLafferty, 1957).

F. Mercaptans

Limited investigations have been carried out upon mercaptans. Here again, fission of the β-bond is favoured over that of the α-bond.

$$R(CH_2)_n\overset{+}{S}H \rightarrow R\cdot + (CH_2)_n\overset{+}{S}H \text{ or } \overset{+}{R} + \text{residue}$$

In some instances hydrogen sulphide is lost. The parent molecular ion is rather abundant which facilitates molecular weight determination (Levy and Stahl, 1957).

G. Aldehydes

Aldehydes also show β-bond fission with concomitant hydrogen migration

$$RCH_2CH_2\overset{+}{C}H_2CHO \rightarrow RC\overset{+}{H}CH_2 + C_2H_4O$$

as the major fragmentation. Again, however, α-fission will occur if there is branching at that carbon atom. The parent molecular ion is fairly abundant for the lower members of this series but, as with many other groups discussed, diminishes rapidly, and may be entirely wanting. The aldehydic group is a basic one in the mass spectrometer and, at moderately high gas pressures, it is easy to observe a $(P+1)^+$ ion.

H. Acids

Among the other important terminal groups there remain carboxylic acids and esters. The former group is relatively little studied since acids have higher boiling points than the corresponding esters and are prone to thermal degradation. The major ions arise by fragmentation at an acyl bond.

$$RC\overset{+}{O}OH \rightarrow R\overset{+}{C}O + \cdot OH$$

When there is a methylene adjacent to the carboxylic group, a β-bond fission occurs with hydrogen migration,

$$RC\overset{+}{H}CH_2 \wr CH_2CO_2H \longrightarrow RC\overset{+}{H}{=}CH_2 + CH_3CO_2H$$

$$\text{(H)} \qquad\qquad \text{or } RC_2H_3 + CH_3\overset{+}{C}O_2H$$

The presence of a re-arrangement ion $m/e = 60$ $(CH_3CO_2H^+)$ in the spectrum is useful evidence of the type of grouping shown; chain branching also influences the fragmentation process, but rarely to the extent of interfering with the basic pattern. The preferred elimination of the longer branch at a point of branching in an alkyl chain is also observed, provided that such branches are rather far from the carboxylic acid group.

Under relatively high gas pressures an ion $(P+1)^+$ is obtained. This is usually present when the parent molecular ion is feeble or entirely absent, and is useful for determining the molecular weight (Happ and Stewart, 1952; Hallgren, Ryhage and Stenhagen, 1957; Hallgren, Stenhagen and Ryhage, 1959; Reed and Reid, 1963).

J. Esters

Esters have been rather extensively studied (Ryhage and Stenhagen, 1963). They may be divided into two classes: methyl esters and the higher esters. The methyl esters are easier to interpret because all fragmentations must arise from the carboxylic acid residue. Studies of the methyl esters of long chain acids have shown a series of ions in which the methoxyl group is retained. These are the more abundant ions of the series, corresponding to successive fissions in the alkyl chain:

$$R \wr (CH_2)_n\overset{+}{C}O_2CH_3 \longrightarrow (CH_2)_n\overset{+}{C}O_2CH_3 + \text{residue}$$

A less abundant series which has lost the methoxyl group is also present. Fission of the β-bond with hydrogen re-arrangement occurs to yield

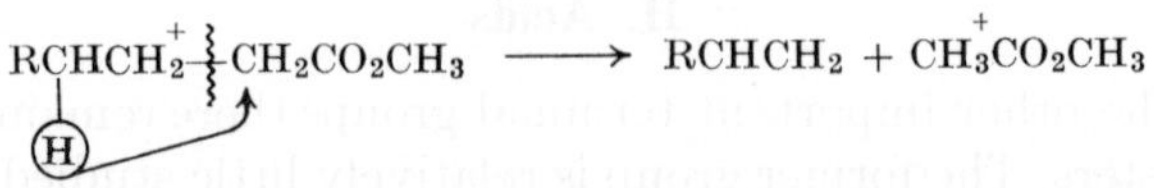

a prominent ion $m/e = 74$, methyl acetate. Moreover, should there be branching on the α-carbon atom then the corresponding methyl ester ($RCH_2CO_2CH_3$) is observed.

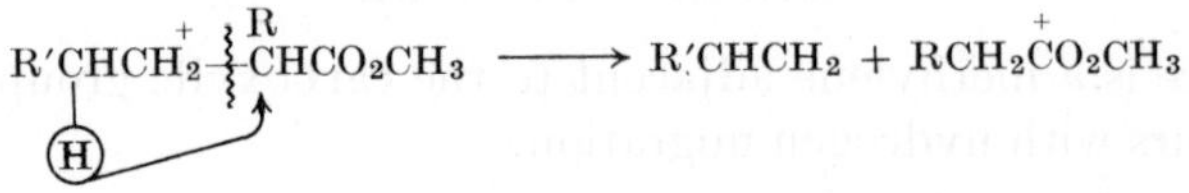

In higher esters, that is those where the methyl has been replaced by an alkyl group with at least two carbon atoms, a further series of ions may be present to complicate the fragmentation patterns. A facile reaction of the type

$$R'CO_2\overset{+}{C}H_2CH_2R \rightarrow R'\overset{+}{C}O_2H + \text{residue,}$$
$$\text{or } R'CO_2H + R\overset{+}{C}HCH_2$$

will occur and the spectrum may contain ions corresponding to the sequential fragmentation of either the acid or the olefin ion.

K. Diesters

Diesters have been little examined; the formation of the ester, and the methyl ester for preference, is usually necessary before any study can be made; the dicarboxylic acids are often too involatile or too unstable for direct analysis. Successful analyses have been made by the use of special techniques (Ryhage and Stenhagen, 1963).

Several series of ions may be expected from the fragmentation of the dimethyl esters. The available evidence shows that the most abundant ions arise from the loss of one methoxyl group.

In this series, as in the monoesters and the carboxylic acids themselves, the parent molecular ion is rather weak, although exceptions do occur.

L. Nitriles and Isonitriles

There remains a series of substituents containing nitrogen. Nitriles give a very feeble parent molecular ion, but the ions $(P-1)^+$ and $(P+1)^+$ are usually abundant. The main fragmentation has already been reported and is

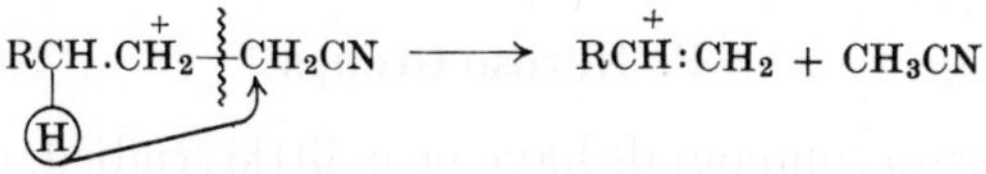

In the special case propionitrile, α-bond fission with elimination of hydrocyanic acid is observed.

Isonitriles, on the other hand, show a pronounced cleavage at the α-bond with elimination of HCN or H_2CN (Gillis and Occolowitz, 1963).

M. Acid Amides

Acid amides show β-fission with hydrogen migration as the main breakdown of the molecular ion which, in general, is of low abundance. This type of re-arrangement is characteristic of most primary acid amides, and it is sometimes observed in secondary amides.

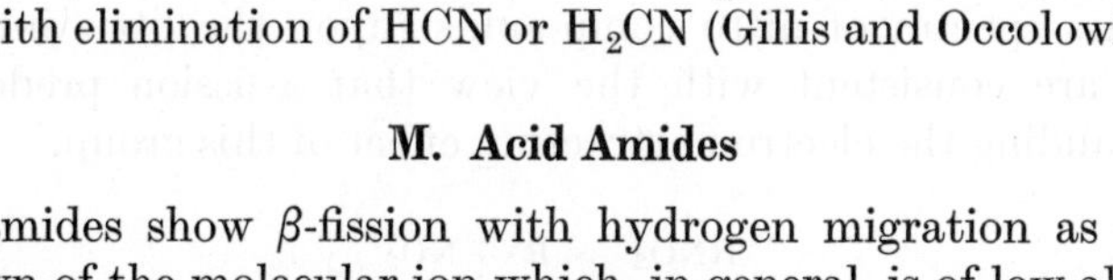

where $R = C_{14}H_{29}$. Secondary amides yield mostly an abundant ion CH_4N^+ arising from the reaction

$$R'CO \overset{+}{|}NHCH_2 \overset{|}{|} CH_2CHR \longrightarrow CH_4\overset{+}{N} + \text{residue}$$

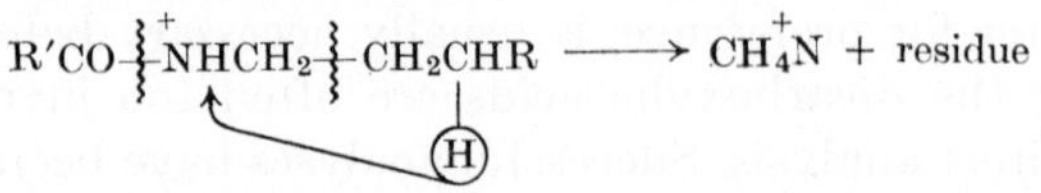

All these types of fragmentation are observed in tertiary amides, each of which must be considered individually (McLafferty, 1956a; Gilpin, 1959; Pelah, Kielczewski, Wilson, Ohashi, Budzikiewicz and Djerassi, 1963).

N. Amines

The principal breakdown of amines has already been mentioned, and it is sufficient to summarize the results. Thus, β-fission with hydrogen re-arrangement is common in primary and secondary amines; tertiary compounds mostly eliminate the longest alkyl chain (Collin, 1952; Collin, 1954).

P. Nitroso Groups

Aliphatic nitroso compounds have been little studied, and the limited evidence available accords well with the principle that α-fission is predominant, with the loss of the alkyl group.

$$R\overset{+}{N}O \rightarrow R \cdot + \overset{+}{N}O$$

Q. Nitro Groups

The mass spectra of nitro compounds (upon the little information reported) are consistent with the view that α-fission predominates, notwithstanding the electron attracting effect of this group.

$$R\overset{+}{N}O_2 \rightarrow R \cdot + \overset{+}{N}O_2$$

The effect does not seem very marked, however, and the influence of other structural characteristics may obscure it. If there is a double bond adjacent to the point of attachment of the nitro group, fission occurs β to the group, in accordance with the usual rules governing the fragmentation of an olefinic compound.

$$\underset{NO_2}{\overset{}{\underset{|}{\Large>}C{=}\overset{+}{C}{-}CH}}{|}R \longrightarrow \Large>C{=}C{-}CHNO_2 + \overset{+}{R}$$

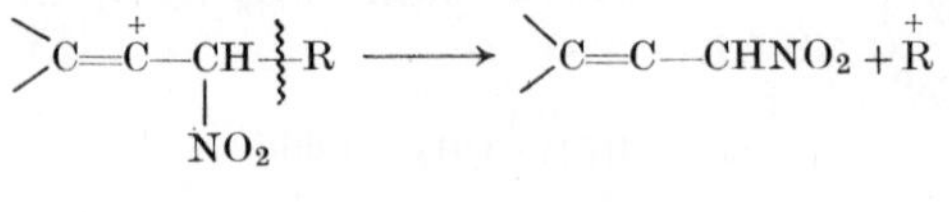

R. Nitrites

Again, nitrites show predominantly α-fission.

$$\overset{+}{RONO} \rightarrow R\cdot + \overset{+}{ONO} \text{ or, more frequently, } \overset{+}{R} + NO_2$$

Occasionally as with isoamyl nitrite fragmentation of the β-bond occurs without a hydrogen re-arrangement.

$$(CH_3)_2CH \wr \overset{+}{C}H_2ONO \longrightarrow CH_2\overset{+}{O}NO + \text{residue}$$

It seems, therefore, that the electron displacement of the nitrite group, in common with the nitro and nitroso groups, is rather unimportant in deciding the point of fission. The β-fission in isoamyl nitrite may, however, be a property of the isopropyl radical contained in the alkyl group, rather than of the nitrite group itself (McLafferty, 1956).

In all three cases discussed, viz. nitroso, nitro, and nitrite substituents, the parent molecular ion is very weak or often absent. Accordingly, it is of little importance in determining the molecular weight.

In addition to the groups already examined, there are some substituents which, being divalent, occur within a carbon chain rather than terminally. These include oxygen in ethers, and sulphur in sulphides.

S. Ethers

Aliphatic ethers of the form ROR′, where neither R nor R′ = Me, have been extensively studied (McLafferty, 1957). If the two substituents are primary alkyl chains, β-fission is favoured in fragmentation.

$$RCH_2\overset{+}{C}H_2 \wr CH_2.OR' \longrightarrow R\overset{+}{C}H{:}CH_2 + ROCH_3$$

If, however, branching is present in the carbon chain, particularly on the carbon atom adjacent to the oxygen, α-fission is preferred. For di-isopropyl ether the predominant fragmentation is

$$(CH_3)_2CH\overset{+}{O}CH(CH_3)_2 \rightarrow (CH_3)_2\overset{+}{C}HO + \text{residue}$$

T. Sulphides

Sulphides show two characteristic features. Fragmentation in the alkyl chains occurs and β-fission is favoured in those sulphides of higher molecular weight which possess at least one large alkyl substituent (Levy and Stahl, 1957).

U. Ketones

Aliphatic ketones, also electron attracting groups, have the characteristic β-fission with hydrogen re-arrangement (Sharkey, Jr., Shultz and Friedel, 1956). Because there are two alkyl substituents in which this type of fission can occur, the spectrum may become complicated by the presence of two daughter molecular ions as a consequence of the reaction sequence

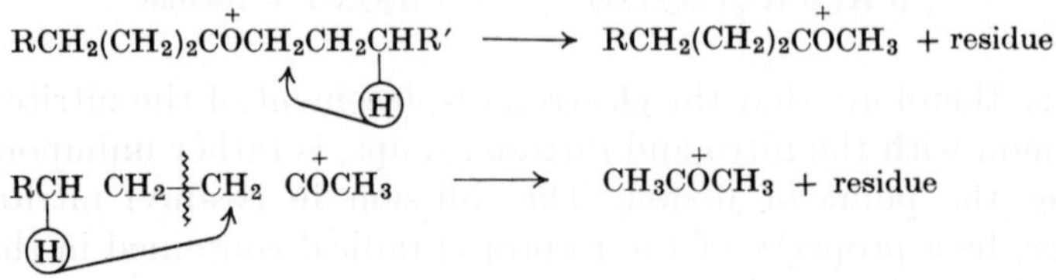

Provided that these complications are recognized, they need not handicap the analysis and, more often, they may aid it, for the presence of acetone in the spectrum necessarily indicates the structure $-CH_2COCH_2-$. The problem is that, since the abundance of the molecular ion of acetone is often very great (even the base peak of the spectrum), this molecule may be considered as an impurity in the substance under examination, a risk which may be minimized by ensuring the purity of the starting material.

Alicyclic ketones belong to this class, and the problem of extensive re-arrangement is, in this instance, overlaid by the characteristic mode of fragmentation of the ring. For cyclohexanone

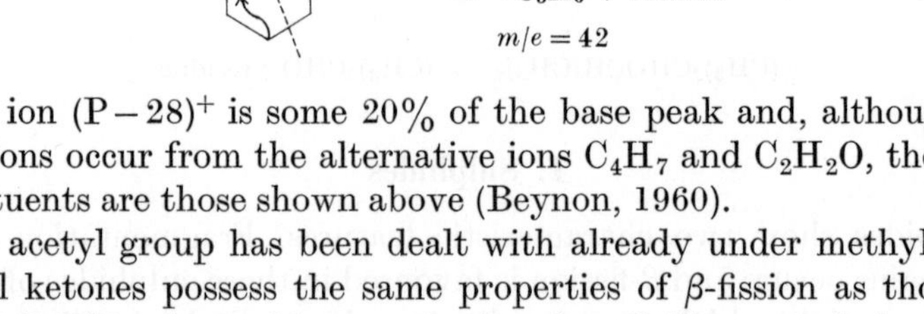

The ion $(P-28)^+$ is some 20% of the base peak and, although contributions occur from the alternative ions C_4H_7 and C_2H_2O, the major constituents are those shown above (Beynon, 1960).

The acetyl group has been dealt with already under methyl esters. Methyl ketones possess the same properties of β-fission as those just recorded. This class of ketone is therefore liable to include acetone as a very abundant re-arrangement ion.

V. Lactones

The lactones are a further group which, so far, remain outside the classification. Two problems here exist: the character of the fragmentation which arises from the electron shifts in the lactone, and the possibility of geometric isomerism with its effect upon fragmentation. The effect of the electron distribution is shown by the characteristic patterns of breakdown of simple lactones. In the case of simple saturated lactones, the fission shown is favoured, with the observed elimination of carbon dioxide (Friedman and Long, 1953).

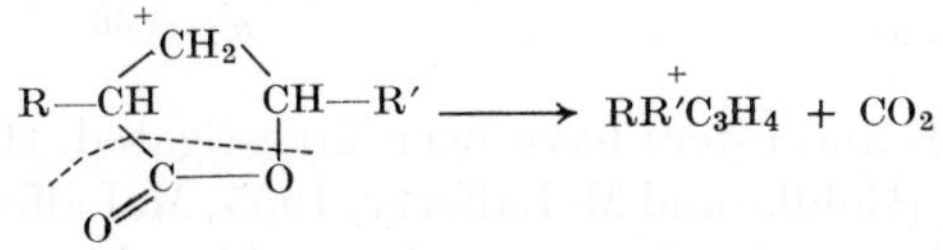

In unsaturated lactones, re-arrangements are often encountered in the formation of fragment ions; for β-angelicalactone a probable fragmentation is

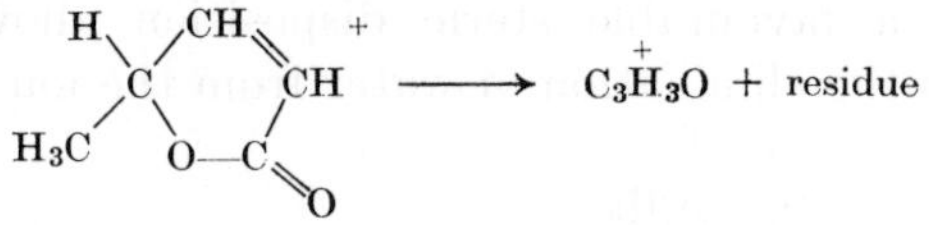

Clearly, the extended conjugation –CH=CH—C=O in the system prevents the normal fragmentation such as is observed in γ-valerolactone above.

W. Aromatic Compounds

Most of the substituents discussed produced distinctive breakdown patterns when directly attached to an aromatic nucleus, the possible exception to this generalization being the halogens. An investigation of a limited group of these shows that the series PhX may be divided into two groups X = F, H and X = Cl, Br and I (Momigny, 1955). The behaviour of fluorine under electron impact is very like that of a hydrogen in benzene. On the other hand, chlorobenzene shows a tendency to yield a phenyl ion, which is more marked when X = Br and I.

$$\overset{+}{PhX} \rightarrow \overset{+}{Ph} + X\cdot.$$

Aromatic alcohols behave in the same way as do aliphatic alcohols. Phenols, however, behave quite differently; they do not fragment significantly, but they do, surprisingly, eliminate carbon monoxide. Methyl ethers eliminate a methoxyl group to a moderate amount at least.

Aromatic aldehydes also tend to lose carbon monoxide which is elided as a neutral fragment. The complete aldehyde group may also occasionally be eliminated, when there is an ion $m/e = 29$ (CHO^+) (Beynon, 1960, p. 361).

$$\overset{+}{\text{PhOH}} \longrightarrow \overset{+}{C_5H_6} + CO$$

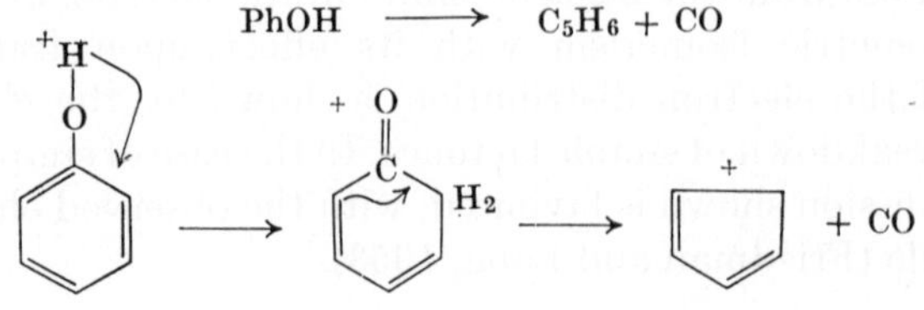

$m/e = 94$ $m/e = 66$

Aromatic acids and esters have been investigated, the former series very extensively (Gohlke and McLafferty, 1955; McLafferty and Gohlke, 1959). In both classes, the characteristic bond breakings in the aliphatic substances are retained; acyl–oxy fission is common in both acids and esters. In the latter group, alkyl–oxygen fission is common amongst the ethyl and higher esters.

Occasionally, a favourable steric disposition allows a particular reaction such as the elimination of water from the ion of *o*-toluic acid.

$$\overset{+}{C_6H_4}\left\langle\begin{matrix}CH_3\\COOH\end{matrix}\right. \longrightarrow \overset{+}{C_8H_6O} + H_2O$$

$m/e = 136$ $m/e = 118$

Aromatic nitrogen-containing compounds, especially where the substituent group is directly attached to the ring, behave quite differently from their aliphatic counterparts. Aromatic primary amines and nitriles generally eliminate hydrocyanic acid.

$$\overset{+}{C_6H_5NH_2} \longrightarrow \overset{+}{C_5H_6} + HCN$$

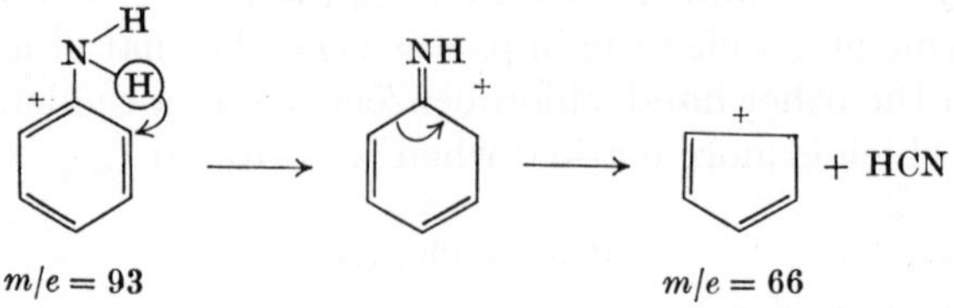

$m/e = 93$ $m/e = 66$

Both nitro and nitroso groups eliminate NO fairly readily, and the former may also lose NO_2. Nitroso compounds, however, unless run at low gas pressures, may dimerize before breakdown, giving rise to ions of mass greater than the true molecular weight.

Mixed alkyl, aryl ketones show a marked tendency to β-bond fission, formally analogous to that occurring in alkylated benzenes. There is, however, no evidence of an associated re-arrangement of the fragment to a tropone ion.

Purely aromatic ketones often eliminate carbon monoxide; fluorenone readily loses carbon monoxide presumably to yield the diphenylene ion.

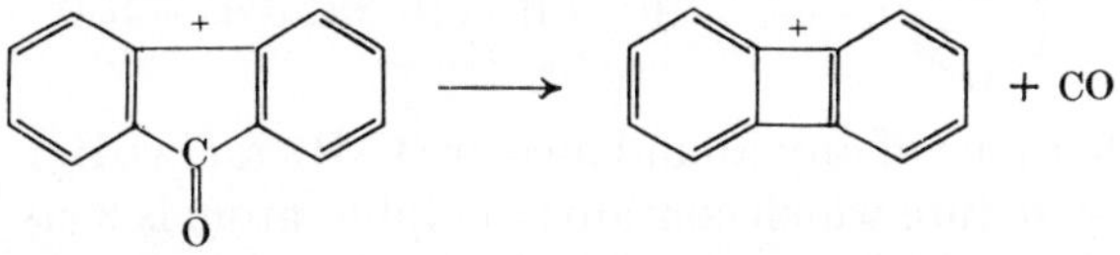

An extremely small amount of fragmentation also occurs within the aryl rings.

Quinones, which have no exact analogy in aliphatic chemistry, eliminate carbon monoxide readily. Thus, anthraquinone loses first one then a second carbon monoxide molecule to yield presumably fluorenone and then the diphenylene ion (Beynon, Lester and Williams, 1959; Beynon, 1960, p. 272).

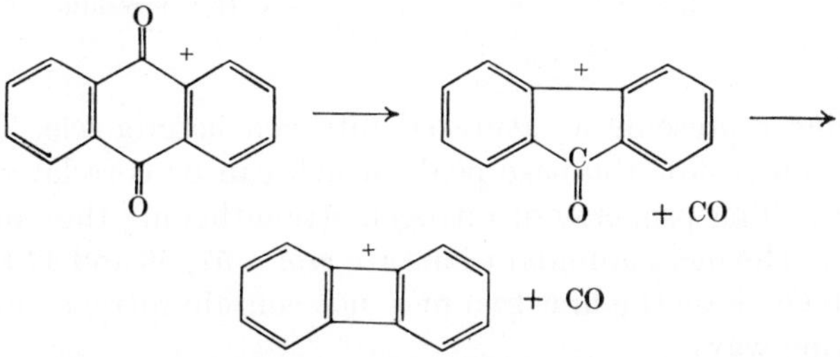

The base peak of the spectrum is usually $(P-28)^{+}$. In all aromatic substances here discussed there is an abundant parent molecular ion which is often the base peak.

X. Cyclic Structures

The third major group, namely cyclic structures which contain a hetero-atom, cover a wide and diverse field of organic compounds. Amongst the many varieties of heterocyclic ring it is not possible to take examples of every class. Accordingly, a few simple examples have been chosen which show the general principles.

Clearly two different categories of compound may be recognized, those in which the heterocyclic atom is part of an alicyclic system, and others in which it is in an aromatic structure. Also, consistent with the other evidence of this chapter, fragmentation is more probable in the alicyclic than in the aromatic series.

Among the cyclic compounds, tetrahydrofuran represents a saturated cyclic ether. The parent molecular ion is rather unstable and is barely present $< 0{\cdot}1\%$. The base peak $m/e = 52$ corresponds to the loss of twenty mass units, which must be the loss of molecular hydrogen and water, probably as follows:

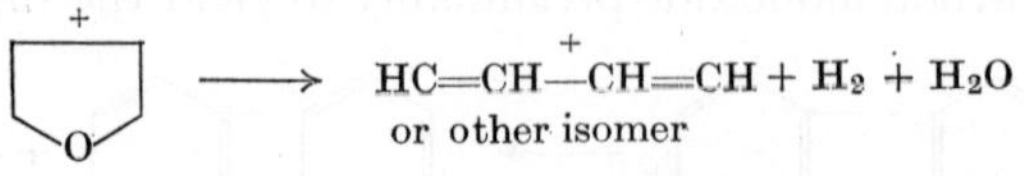

The only other ions of any abundance are $C_2H_5^+$ and $C_2H_3^+$.

A similar structure which contains a sulphur atom is 2-*cis*-5-dimethylthiacyclopentane. The parent molecular ion is quite abundant $m/e = 116$ (38%), whilst the base peak represents the loss of one methyl group. The next most abundant ion $m/e = 59$ (40%) must contain the hetero-atom and is probably the result of a β-bond fission followed by fragmentation at a point of branching, all accompanied by a hydrogen migration

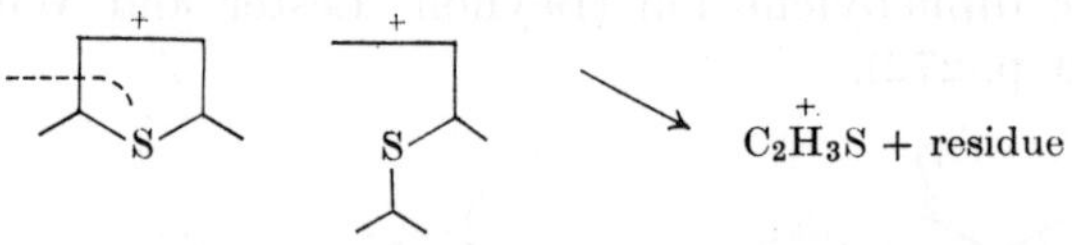

Piperidine represents a saturated nitrogen heterocycle. The parent molecular ion is now the base peak, which can be correlated with the electron repelling property of nitrogen strengthening the bonds within the system. The most abundant ions are $m/e = 57$, 56 and 44 (52%, 51% and 40%). Of these the first two may not contain nitrogen and arise in the following way

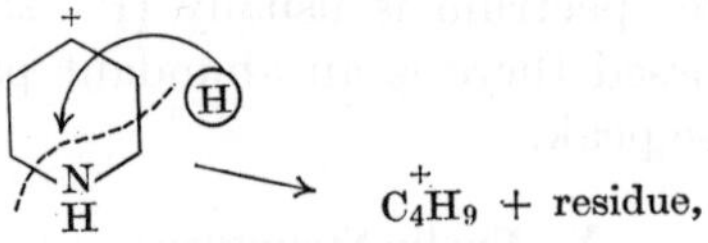

or else, $C_4^+H_8$ +residue if there is not a concomitant hydrogen migration. Alternatively, fragments which do include nitrogen may be formed by the appropriate fissions. The $m/e = 44$ most probably does contain the nitrogen and is C_2H_6N. The method of fragmentation is necessarily speculative, but probably follows the route

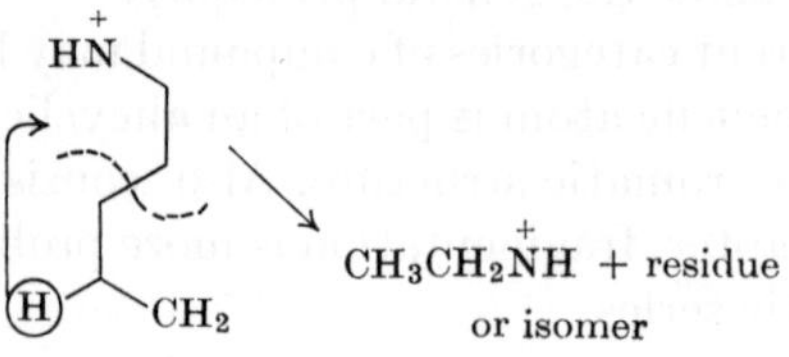

Compounds which contain two hetero-atoms are 1,3-dioxalone and *N*-methylmorpholine. In the former there is a small parent molecular ion $m/e = 74$ (6·0%) and the base peak corresponds to the loss of one hydrogen atom; further prominent ions occur at 43^+ (70%), 44^+ (60%), and 29^+ (57%).

Ether oxygens are usually associated with β-bond fission. In the particular instance of the inter-oxygen methylene, the hydrogens are bonded β to each oxygen. It is therefore reasonable to suppose that the hydrogen lost comes from here.

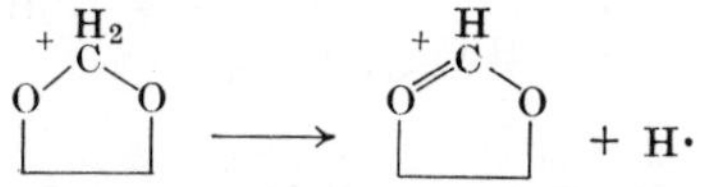

Whether this is indeed the case, or whether the dioxalone first protonates and then eliminates molecular hydrogen has not been decided. In any event the appearance of a $(P-1)^+$ ion is characteristic of the system. The ion $m/e = 73$ will be rather stable as it can exist in two oxonium forms, one of which is shown above.

The two ions 43^+ and 44^+ have the formulae $C_2H_3O^+$ and most probably $C_2H_4O^+$, and these are readily obtained by the bond cleavage

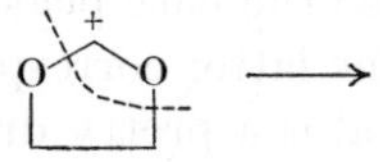

$\overset{+}{C_2H_4O}$ + CH_2O without, and
$\overset{+}{C_2H_3O}$ + CH_3O (or more likely $\overset{+}{C_2H_3O} + HCO + H_2$)
with a hydrogen migration.

If the last, energetically more favourable, reaction sequence is the one followed, then the ion $m/e = 29$ (57%) would have the formula CHO^+ and would arise from an alternative ionization. The likely mechanism for the former fragmentation is as follows:

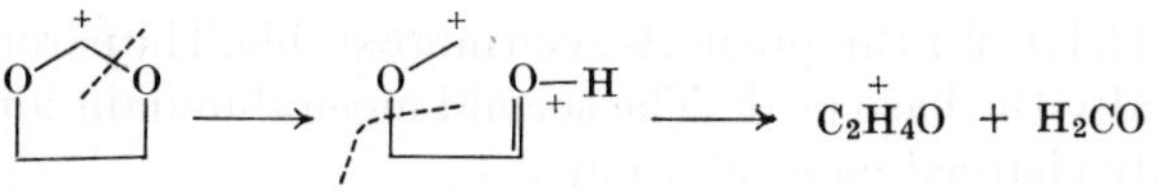

$m/e = 74$ $m/e = 44$

In *N*-methylmorpholine, the fragmentation pattern is more complex. The parent molecular ion is rather abundant (25%), and $m/e = 43$, corresponding to the loss of fifty-eight mass units, is the base peak. The other abundant ion 42^+ (49%) relates to the formation of the ion 43^+.

A fission consistent with the production of the ion $m/e = 42$ without and 43^+ with re-arrangement is as follows:

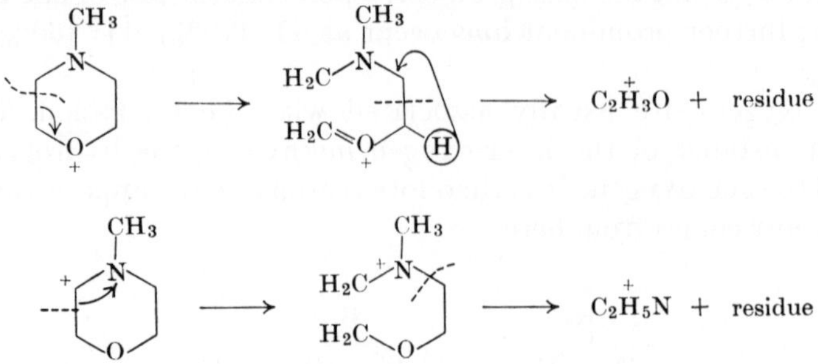

Y. Heterocyclic Compounds

Heterocyclic compounds which have an affinity with the aromatic hydrocarbons are much more stable. The parent molecular ion is nearly always the base peak. The presence of a basic hetero-atom, particularly nitrogen, does not markedly lower the stability of the parent molecular ion but, because it increases the basicity, the formation of the doubly charged ion is more probable, and now it may be the second most abundant ion in the spectrum.

In diphenylene oxide (dibenzofuran), the three most abundant ions are the parent molecular ion $m/e = 168$ which is also the base peak, the doubly charged ion 84^+ (11%), and 139^+ (23%). The latter corresponds to the elimination of the formyl radical and indicates a pretty drastic re-arrangement within the fragment ion so formed. The shape of this hydrocarbon ion cannot be proved. It may contain three rings, six-, four-, and five-membered, but this is mere speculation.

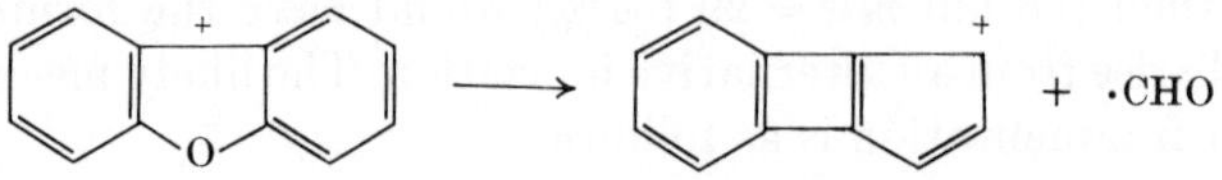

A more elaborate example, a sulphur-containing aromatic compound, dinaphtho-(2,1,1′,2′)-thiophene, is even more stable. The parent molecular ion 284^+ is also the base peak. The second most abundant ion 142 (15%) is the doubly charged molecular ion

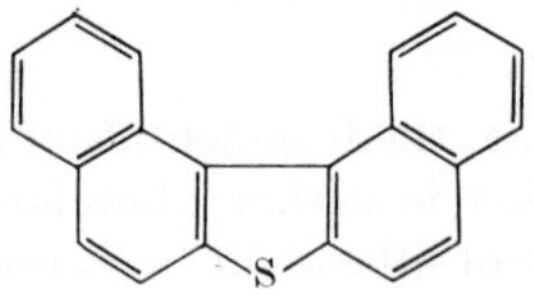

Dinaphtho-(2,1,1′,2′)-thiophene

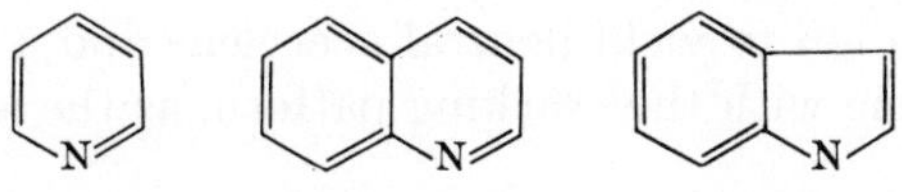

Pyridine Quinoline Indole

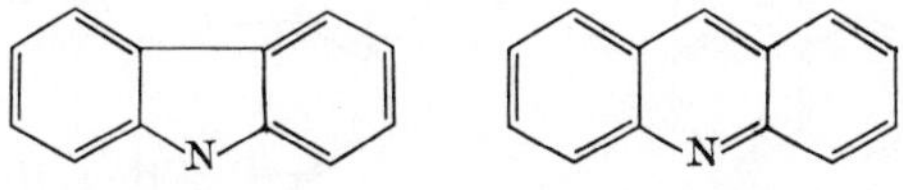

Carbazole Acridine

A much larger group of nitrogen-containing aromatics has been examined because of the importance of these compounds in alkaloid chemistry. In pyridine, quinoline, indole, carbazole, and acridine, the base peak is also the parent molecular ion. Acridine and carbazole are simple in that the only other abundant ion is $(P-1)^+$, which amounts to some 14% of the base peak in each instance. The remaining three all lose twenty-seven mass units as hydrocyanic acid, an elimination which has previously been remarked in aniline and other arylamines.

This loss must in all cases be accompanied by some re-arrangement which might yield a new conjugated aromatic hydrocarbon or even an open chain alkyne; for pyridine one could conceivably obtain either cyclobutadiene or vinylacetylene.

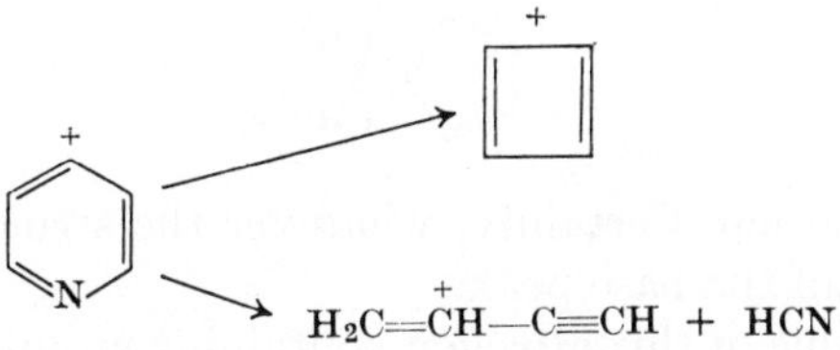

It is imprudent to speculate much upon this. However, the high abundance $m/e = 52$ (71%) suggests that the ion must be rather stable, which would favour vinylacetylene rather than cyclobutadiene. The latter might be expected to undergo a facile depolymerization to yield acetylene, both as the ion and the neutral molecule.

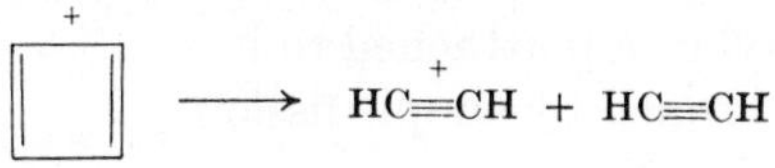

This supposition seems confirmed by observations upon quinoline, which may yield either a substituted cyclobutene or an alkyne.
In the first of these, one might expect the benzo-substituted compound to be more stable than the corresponding ion obtained from pyridine, which is not the case. The second may further dissociate by loss from

the intermediate ion to yield neutral acetylene and a phenyl ion. This seems more in line with the cracking pattern, as the ion corresponding

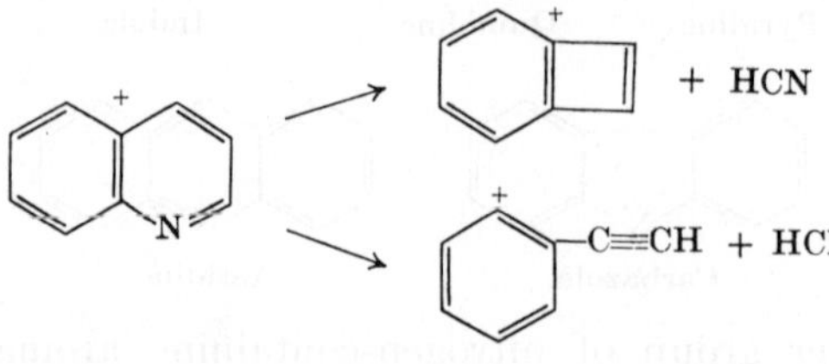

to the loss of hydrogen cyanide (102^+) is not very abundant (23%). With indole, the phenylacetylene cannot be formed. There is one hydrogen too few for the loss of HCN accompanied by re-arrangement to yield a stable tropylium ion.

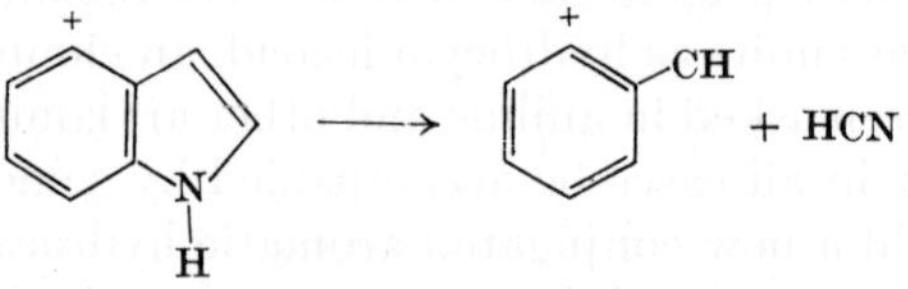

The alternative is a ring reduction process to give

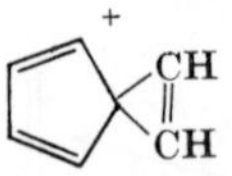

a spiroheptatriene ion. Certainly, whatever the structure of the ion it amounts to 40% of the base peak.

The final example of this nitrogen-containing group is nicotine, which has as principal ions the following: $m/e = 162$ ($C_{10}H_{14}N_2^+$; 18%) the parent, 133 ($C_9H_{11}N^+$; 27%), 84 ($C_5H_{10}N^+$) the base peak, 42 (20%), and 28 (17%). The rather abundant parent molecular ion is consistent with the largely aromatic character of the molecule. The loss of twenty-nine mass units may or may not include the loss of nitrogen. If a nitrogen is eliminated (as is most likely), it is from the alicyclic ring and is accompanied by the methyl group attached to it.

The base peak is formed by simple fission

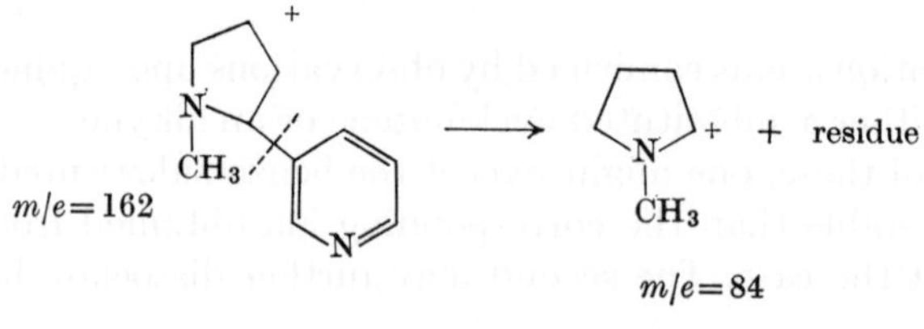

Mass 42^+ is probably derived from this by further cleavage.

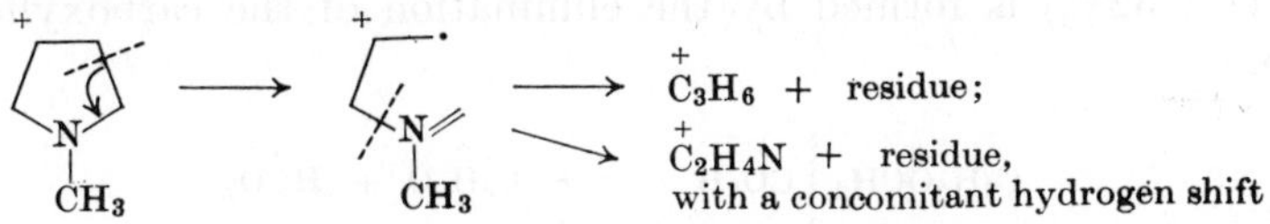

There is some contribution from the doubly charged ion. Exact mass measurements confirm that the major ionic species present is $C_2H_4N^+$. The remaining ion $m/e = 28$ will be obtained by fragmentation and concomitant migration, the smaller mass bearing the charge.

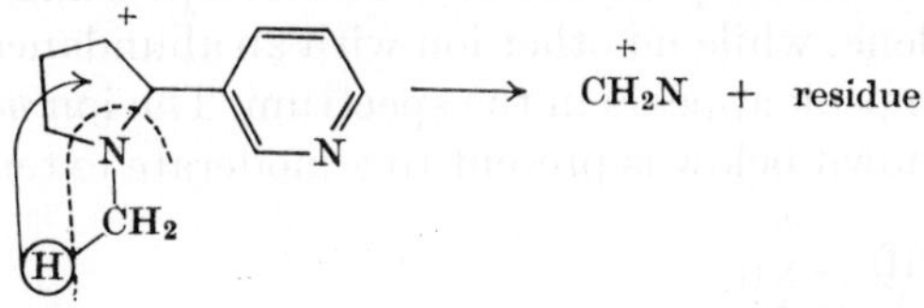

A spectrum from a double-focusing instrument is not much help (Chapter 5, p. 123).

Ethane dithiol has a rather more stable molecular ion (64%). Again the base peak 47^+ corresponds to the fission

$$HSCH_2 \wr \overset{+}{C}H_2SH \longrightarrow \overset{+}{C}H_3S + CH_3S\cdot$$

as might be predicted from a consideration of inductive effects. Other ions are present also: 61^+ ($C_2H_5S^+$; 59%) which corresponds to the loss of HS; 60^+ (69%) which results from the elimination of hydrogen sulphide; 27^+ (49%) which must be $C_2H_3^+$ and arise by a complex re-arrangement; 59^+ (34%) formed in a like manner

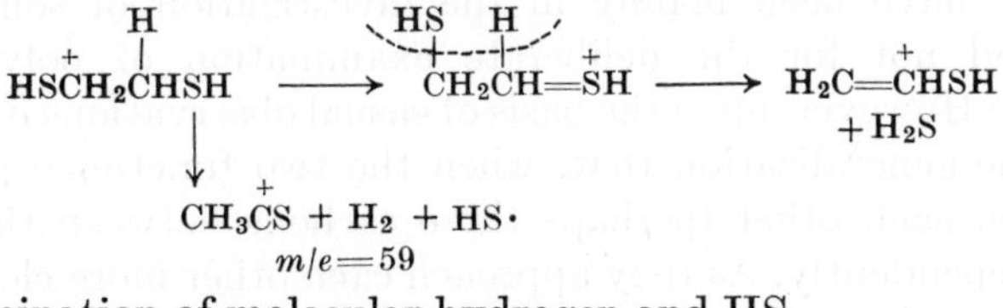

by the elimination of molecular hydrogen and HS·.

A molecule which has rather different functional groups in juxta-position is ethoxyacetic acid. There is no parent molecular ion. The base peak is at $m/e = 31$, corresponding to the formation of a methoxyl ion which must be formed by more than one fission and probably by an extensive, associated re-arrangement

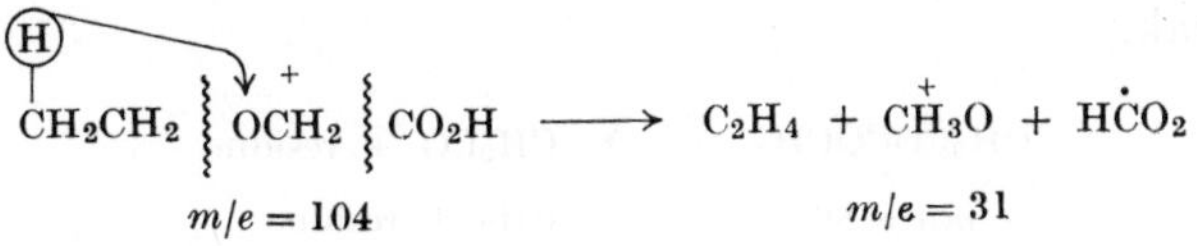

or some similar process. The other abundant ion of high mass $m/e = 59$ ($C_3H_7O^+$; 52%) is formed by the elimination of the carboxylic acid group.

$$\overset{+}{C_2H_5OCH_2} \,\wr\, CO_2H \longrightarrow \overset{+}{C_3H_7O} + H\dot{C}O_2$$

Amongst the lower mass ions 29^+ (62%) corresponds to the removal of an ethyl group and 18^+ (61%) to the elimination of water.

Finally, the cracking pattern of 2-aminoisopropanol shows no parent molecular ion but a base peak at $m/e = 58$ corresponding to the loss of a hydroxymethylene, while no other ion with an abundance of as much as 20% of the base peak appears in the spectrum. The ion $m/e = 41$ arising by the fission shown below is present to a moderate extent (18%).

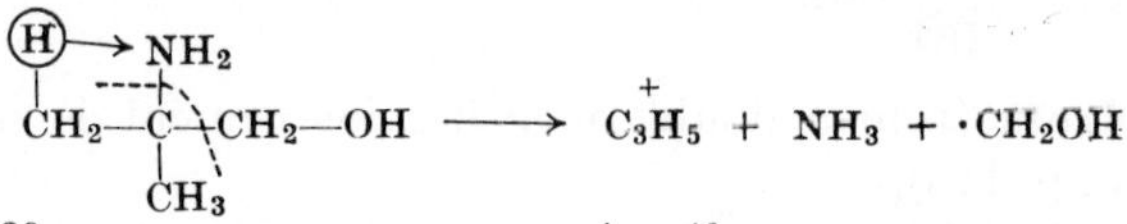

$m/e = 89$ $m/e = 41$

In all these examples, the proximity of the functional groups has led either to fissions not observed in monofunctional molecules, or to the emphasizing of fragmentations which normally occur only slightly.

Z. Polyfunctional Substances

Relatively little systematic study of molecules having more than one functional group has been reported. When such observations have been made, they have been mainly in the investigation of some natural product and not for the deliberate examination of polyfunctional compounds. However, upon the basis of casual observations it is possible to make the generalization that, when the two functional groups are remote from each other (perhaps three carbons between them), they behave independently. As they approach each other more closely, some interaction may be observed. This effect has been previously reported in the discussion upon dialkenes.

The acetyl group is rather stable, and biacetyl may be expected to break into two equal halves. Therefore, the parent molecular ion should not be very abundant; this is indeed the case. The acetyl ion $m/e = 43$ is also the base peak. Methyl ions appear in a significant amount (33%) of the base peak.

$$\underset{m/e\,=\,86}{\overset{+}{CH_3COCOCH_3}} \longrightarrow \overset{+}{CH_3CO} + \text{residue}$$

$$\qquad\qquad\qquad \searrow \overset{+}{CH_3} + \text{residue}$$

In the case of acetonylacetone, the ion $m/e = 43$ is also the base peak. There is a feeble parent molecular ion (3%), and the others present correspond to the loss of a methyl radical, or the formation of the methyl ion

$$CH_3CO\overset{+}{C}H_2CH_2COCH_3 \longrightarrow \overset{+}{C}H_3CO + CH_3CO\dot{C}_2H_4$$
$$m/e = 114 \qquad\qquad m/e = 43$$

The parent molecular ion of ethylene glycol monomethyl ether is somewhat more abundant (6%), but even so the base peak $m/e = 45$ corresponds to the loss of thirty-one mass units. This is to be expected since the combined, strong inductive effects of the hydroxyl and methoxyl groups will greatly weaken the bond joining them. Then one has either the fission

$$CH_3O \,\wr\, CH_2\overset{+}{C}H_2OH \longrightarrow CH_3O\cdot + C_2\overset{+}{H}_5O$$

or

$$CH_3\overset{+}{O}CH_2 \,\wr\, CH_2OH \longrightarrow CH_3\overset{+}{C}OCH_2 + \cdot CH_2OH$$

Both are possible and are so far indistinguishable. On a consideration of inductive effects, the latter mechanism is to be preferred.

CHAPTER 5

The Mass Spectrometry of Natural Products

The application of mass spectrometry to natural products is a relatively new development. Whilst occasional studies had been made upon particular terpenes about six years ago (Friedman and Wolf, 1958), and steroids examined about the same time (Reed, 1958; Friedland *et al.*, 1959), it is only comparatively recently that these subjects have been systematically examined (Biemann, 1962b). The field is very wide and rather ill-defined, so that a somewhat arbitrary selection of topics must be made. Those chosen have been selected not only because they are representatives of rather large fields in natural product chemistry, but also because they have some bearing upon the systematic treatment of cracking patterns. The main sections discussed are therefore steroids, terpenes (Reed, 1963), polyhydric alcohols and alkaloids Antonaccio *et al.*, 1962; Budzikiewicz, Wilson and Djerassi, 1962; Djerassi, 1963; Djerassi *et al.*, 1962a–i; Gilbert *et al.*, 1962a,b, 1963; Lund *et al.*, 1963a,b; Lynch *et al.*, 1963; Nakagawar *et al.*, 1962; Olivier *et al.*, 1963; Plat *et al.*, 1962a,b,c; Sandoval *et al.*, 1962; Shapiro, Wilson and Djerassi, 1963.

A. Acyclic Terpenes

The terpenes are all based upon the polymerization of an isoprene unit. Isoprene has a rather abundant parent molecular ion, and the base peak corresponds to the loss of a single hydrogen atom. Apart from this ion the other abundant ions are 53^+ (86·2%) and 39^+ (71%). All these fragmentations are shown in the following composite diagram.

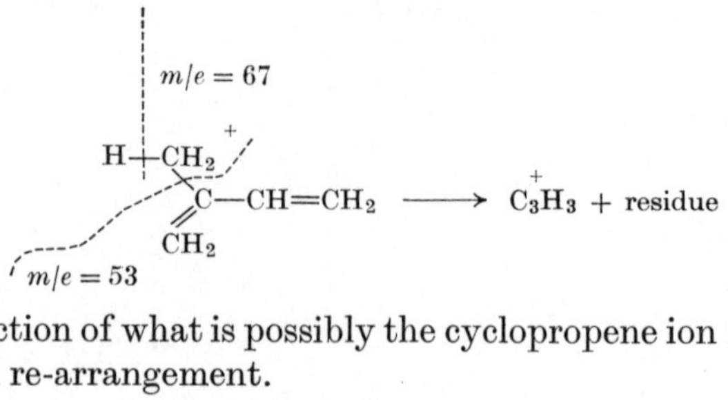

The production of what is possibly the cyclopropene ion must involve some skeletal re-arrangement.

Dimerization of isoprene will lead, formally at any rate, to the monoterpenes, which contain ten carbon atoms. These may be acyclic, monocyclic or possess two rings. The acyclic members of the series are represented by myrcene and allo-ocimene. They contain three double bonds and each terminates in an isopropenyl group. Myrcene has a single bond which is doubly allylic, a feature that correlates well with the low abundance of the parent molecular ion (8%). This is absent in allo-ocimene and accordingly the molecular ion 136^+ is more abundant (41%).

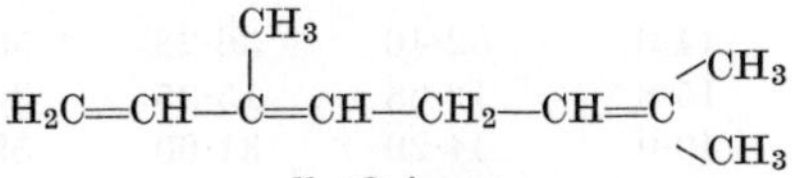

allo-Ocimene

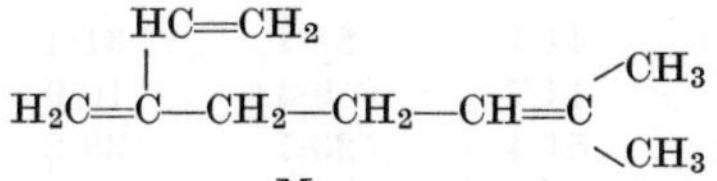

Myrcene

The base peak in myrcene corresponds to the ion $m/e = 41$, even though this cannot be derived readily without re-arrangement or at least extensive bond migration. In allo-ocimene, on the other hand, the base peak is at $m/e = 121$, corresponding to the loss of a methyl group. The three methyls present are attached vinylically to double bonds. Even in such circumstances the methyl is known to be lost (A. P. I. project 44).

B. Cyclic Terpenes

A more extensive series of cyclic monoterpenes has been examined including camphene, the isomeric pinenes, pinane, fenchene, and the menthadienes one of which α-1,8(9)-*p*-menthadiene is better known as d-limonene; *p*-menthene has also been examined. The principal ions in the group are shown in Table XXII. The base peak 93^+ in the case of the pinenes, camphene and α-fenchene is clearly the loss of the disubstituted bridge carbon. Thus, for camphene one has the sequence

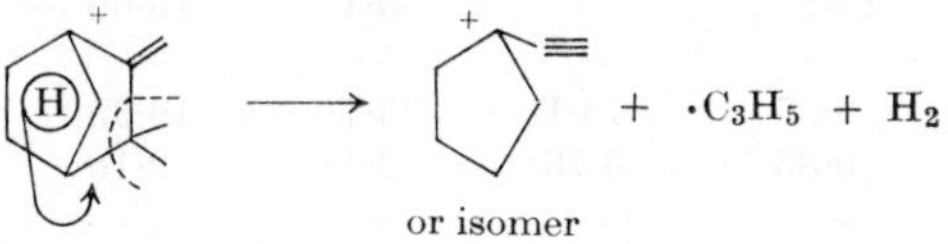

One obvious exception is pinane ($C_{10}H_{18}$), which has two more hydrogens than the others. Therefore the loss of the central *gem*-dimethyl system will yield an ion $m/e = 95$ (74%), which is second in abundance to the base peak. The proposed mechanism here is

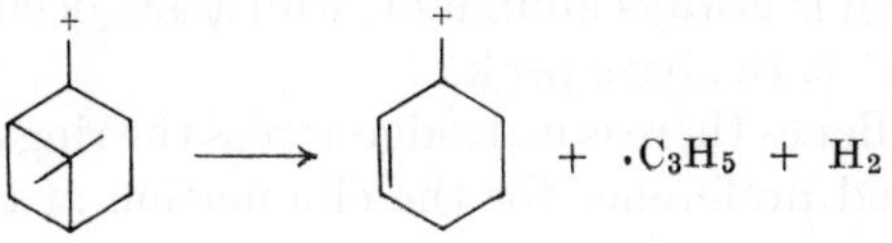

It will be noticed that the neutral fragments proposed are molecular hydrogen and propenyl rather than a propyl radical. This interpretation is preferred since the ion 41^+ which may be formed in an alternative

TABLE XXII

m/e	I	II	III	IV	V	VI	VII
27	44·1	21·1	31·4	44·0	32·46	36·28	58·72
29	14·7	9·44	10·9	15·8	12·68	15·25	34·11
39	51·4	23·7	33·2	49·0	44·29	31·60	59·17
40	10·4	—	—	—	12·02	—	9·75
41	58·6	23·2	63·9	58·9	34·68	42·24	100·0
43	—	—	—	—	—	11·79	16·51
53	15·3	10·5	14·0	21·4	28·23	18·27	24·29
55	—	—	—	—	—	25·71	86·24
67	33·7	—	—	—	40·19	40·19	49·35
68	24·5	—	—	—	100·0	55·06	33·72
69	—	—	46·7	—	—	16·81	49·29
77	23·0	22·1	18·3	30·9	15·87	11·12	—
79	37·5	17·7	19·9	62·6	25·18	14·98	10·50
80	12·2	9·81	10·4	47·0	10·18	—	—
81	—	—	—	27·7	10·54	26·22	44·90
82	—	—	—	—	—	14·96	57·24
83	—	—	—	—	—	—	53·88
91	21·8	21·2	13·2	26·2	14·52	—	—
92	—	29·7	—	16·1	16·33	—	—
93	100·0	100·0	100·0	100·0	53·36	—	—
94	16·7	—	13·5	30·3	18·77	10·19	—
95	21·9	—	—	—	—	100·0	74·42
96	—	—	—	—	—	16·48	29·13
107	29·2	—	—	25·3	14·48	—	—
121	62·6	13·2	—	39·1	16·69	—	—
123	—	—	—	—	—	13·52	19·96
136	14·2	8·05	7·01	23·7	19·35	—	—
137	1·5	0·85	0·76	2·58	2·16	—	—
138	—	—	—	—	—	26·73	4·52
139	—	—	—	—	—	3·46	0·65

I = Camphene, II = α-pinene, III = β-pinene, IV = α-fenchene, V = d-limonene, VI = 1-*p*-methene, VII = pinane.

charge distribution is always abundant, whilst 43^+ occurs only in pinane and even then 41^+ is the base peak.

In the menthadienes there is no bridge across the ring and, accordingly, there is no marked preference for the elimination of the *gem*-dimethyl system, which is now present as an isopropyl. Consequently, $m/e = 93$ is the base peak only in 1,4-*p*-menthadiene, although it is very abundant in the others. 1-*p*-Menthene has one less double bond and gives the base peak at $m/e = 95$. It is not always easy to see why the isopropyl is retained or lost. In the case of 1,4-*p*-menthadiene, it is necessary to suppose double bond migration in order to make this observation consistent with the numerous others. The sequence is, therefore,

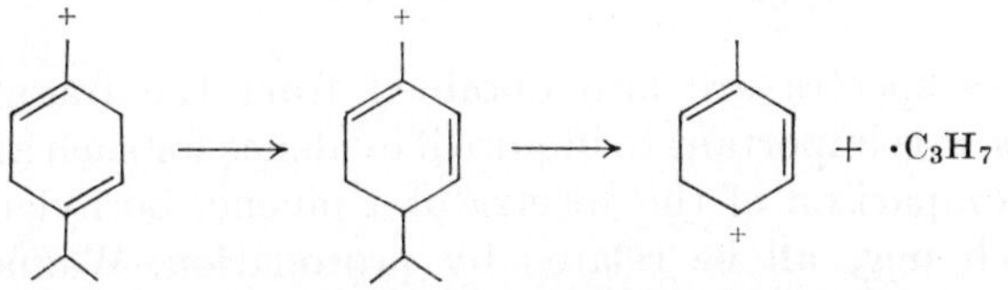

the last step being allylic fission. Having deduced such a system it is easy to provide a *rationale* for it; a conjugated double bond system will be more stable than a non-conjugated system. Unfortunately, unless there is some other feature that suggests the molecular structure, these niceties will not be suspected.

A further interesting point concerning fragmentation arises in the spectrum of α-1,8(9)-*p*-menthadiene, namely the formation of the ion $m/e = 68$, which is also the base peak of the spectrum.

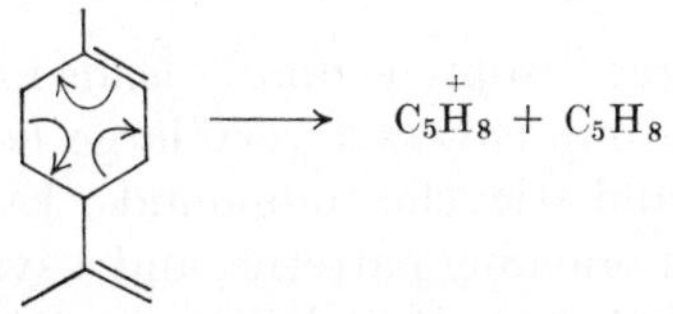

One possible mode of formation is by a retro-Diels-Alder reaction, either on the ion (in which case these products are obtained directly) or in the heated inlet system through which the material is introduced, when isoprene is formed and subsequently ionized. An alternative means of fission is by rupture of two of the allylic bonds present in the molecular ion.

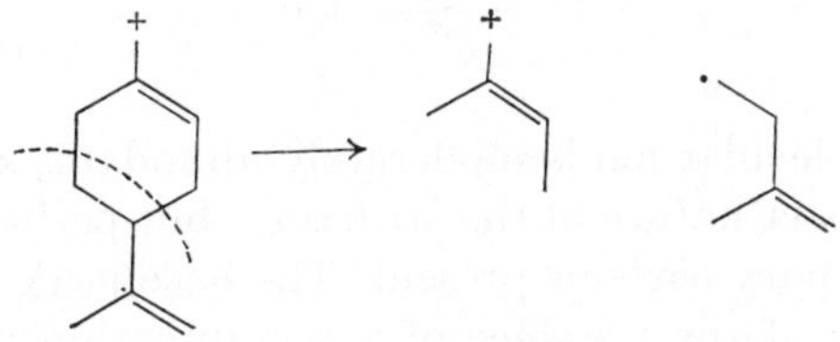

The two mechanisms cannot be distinguished upon the present evidence.

The second point is of considerable theoretical interest. It is well known that many terpenes undergo Wagner-Meerwein re-arrangements. Thus, α-pinene on treatment with hydrogen chloride readily re-arranges to give bornyl chloride as the final product.

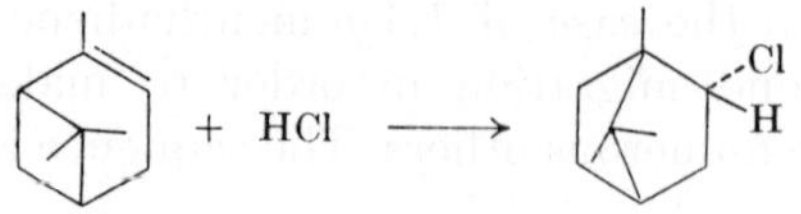

Since mass spectra are also obtained from the fragmentation of positive ions, it is important to inquire if evidence for such arrangements do occur. Comparison of the spectra of α-pinene, bornylene and camphene (which may all be related by protonation, Wagner-Meerwein re-arrangements and deprotonation) suggests that no such re-arrangements are to be observed. Although the spectra show similarities which, having regard to the structures, may be expected, there are certain differences that seem to exclude the idea as a working hypothesis. The base peak of the spectrum of α-pinene is at 93^+, that of camphene also. Camphene has an ion $m/e = 68$ (24·5%) that is absent in the other spectra.

C. The Sesquiterpenes

Condensation of three "isoprene units" leads formally to a sesquiterpene ($C_{15}H_{24}$). This group covers a very large field, including acyclic, monocyclic, bicyclic and tricyclic compounds. Even so, comparatively little is known of their cracking patterns, and a systematic study of this group has only recently begun (H. C. Hill, personal communication). One complex example, longifolene, must suffice.

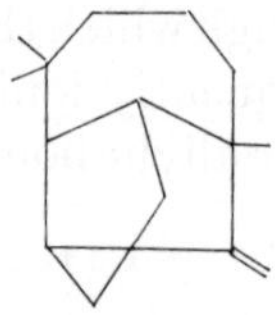

The parent molecular ion is moderately abundant, surprisingly so in view of the compact nature of the structure, but perhaps related to the numerous quaternary carbons present. The base peak $m/e = 41$ may be expected in view of the presence of a *gem*-dimethyl group. The other main ion which has an abundance of half the base peak is $m/e = 91$ (50%) and corresponds to $C_7H_7^+$. It is not easy to decide the origin of this fragment.

D. The Diterpenes

The polymerization of four isoprene units would lead to a group of $C_{20}H_{32}$ molecules. A series is known which may be so derived theoretically, and its members are known as diterpenes. Most analyses have been carried out on the tetracyclic diterpenes of which phyllocladene is a suitable example (Kelly, Reed and Reid, 1962).

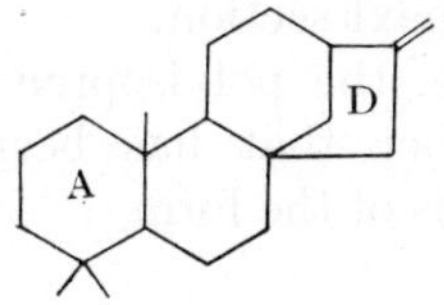

The parent molecular ion is also the base peak of the spectrum confirming the stable nature of the tetracyclic system. Further abundant ions $m/e = 257$ (62%) and 229 (63%) correspond to the loss of a methyl from the *gem*-dimethyl and, probably, forty-three mass units from either ring A or ring D.

Other abundant ions $m/e = 55$ ($C_4H_7^+$; 66%), 119 ($C_9H_{11}^+$; 62%) and 105 ($C_8H_9^+$; 72%), probably arise by the following fragmentations

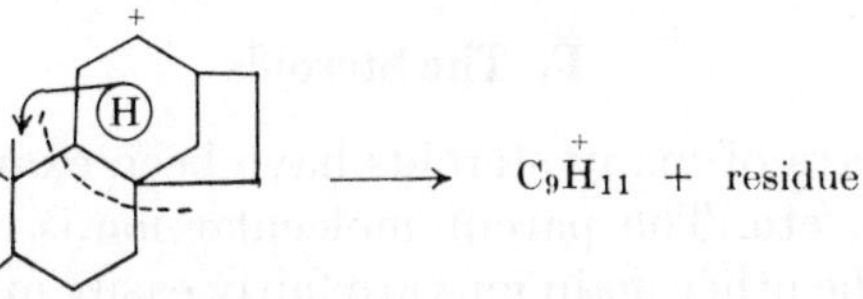

but the formation of the others is not at all obvious.

E. The Triterpenes

The next abundant series of terpenes refers to the tetra- and pentacyclic triterpenes. An example of one type is $\Delta^{9(11)}$-lanostene and of the other α-amyrene.

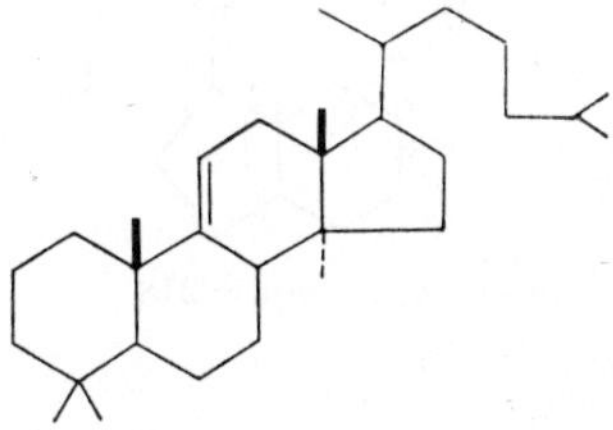

$\Delta^{9(11)}$-Lanostene

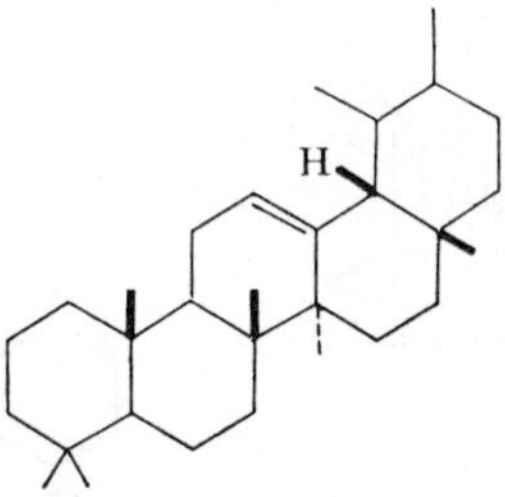

α-Amyrene

Notwithstanding the different molecular structure, the superficial resemblance in the cracking patterns is very striking. In both cases, the parent molecular ion is roughly 10% of the base peak, and is formed in each instance by the loss of fifteen mass units. This must refer to the loss of one methyl from the *gem*-dimethyl group. The next largest ion, about 27% for each compound, is $m/e = 95$ (C_7H_{11}). The most extensive studies amongst the triterpenes are concerned with oxygenated derivatives and these are discussed in the next section.

The remaining terpenes, the polyisoprenoids, have been but little studied. One such member that has been examined is solanesol, $C_{45}H_{73}OH$. The structure is of the form

$$\mathrm{H.CH_2{-}\underset{\displaystyle CH_3}{\underset{|}{C}}{=}CH{-}CH_2{-}(CH_2{-}\underset{\displaystyle CH_3}{\underset{|}{C}}{=}CH{-}CH_2)_7.CH_2{-}\underset{\displaystyle CH_3}{\underset{|}{C}}{=}CH.CH_2OH}$$

and this is revealed by the mass spectrum in which the loss of five carbon fragments is associated with a more abundant ion, corresponding to fission at the allylic carbon–carbon bonds. The methyl groups are vinylically disposed and, accordingly, there is no marked loss of a methyl group.

F. The Steroids

The mass spectra of many steroids have been examined, also androstanes, cholanes, etc. The parent molecular ion is nearly always very abundant, and the other main ions are fairly easily interpreted. The loss

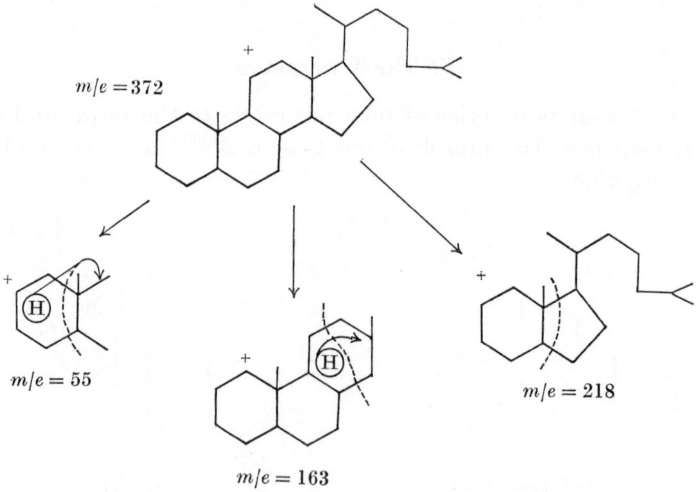

of one methyl group is often facile, and there is further evidence for the loss of part of ring A, A and part of B, and ring D with the entire side chain. This last is a particularly abundant ion and is very useful in determining the number of carbon atoms in the side chain. The fragmentations as applied to cholestane are shown here.

G. Oxygenated Compounds

There is a wide variety of oxygenated compounds among the terpenes, including alcohols, ketones, aldehydes, and ethers. In general, the introduction of the substituent group modifies the mass spectrum obtained in much the same way as was discussed for substituted alkanes and cycloalkanes. For example, the mass spectrum of camphor shows a rather abundant parent molecular ion (29%) with a base peak 95^+ ($C_7H_{11}^+$) and three fairly prominent ions $m/e = 41$ (94%), 81 (75%) and 39 (60%). These ions probably arise as follows:

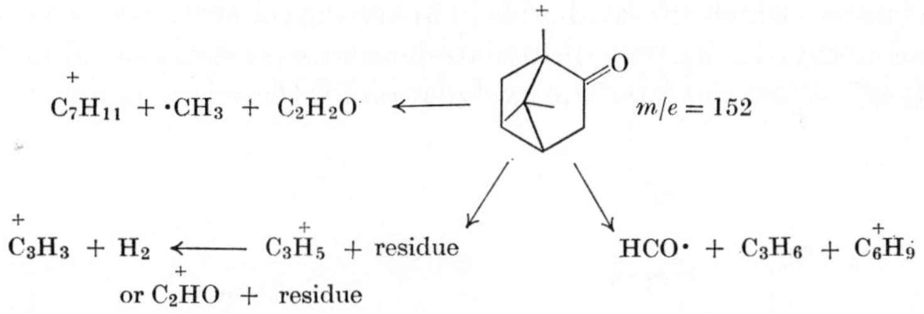

Among the higher terpenes there has been some study of diterpene lactones; one such is rosenonolactone. This simple example does allow of some analysis, a process not always possible unless mass defect measurements have been used to determine the composition of fragment ions. The main ion in the mass spectrum, apart from the parent molecular ion, occurs at $m/e = 165$ ($C_{11}H_{17}O^+$).

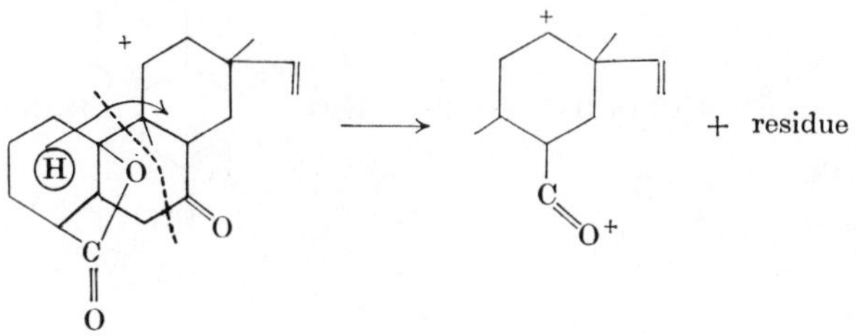

The presence of a substituent in a steroid usually does not alter the mode of fragmentation entirely. Most steroids lose part of ring A to yield two ions, one of which is $m/e = 55$, formed by the fission

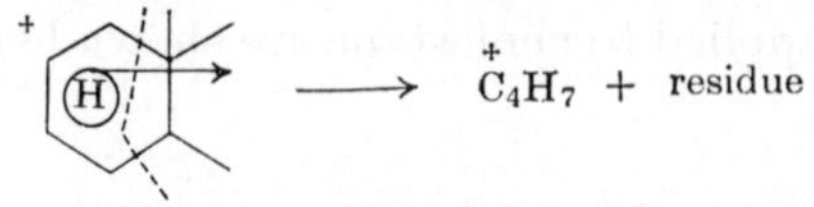

The other yields a radical, presumably $C_4H_7\cdot$ and an ion $(P-55)^+$.

A similar fission occurs in the case of 3-ketosteroids, except in these the ion is $m/e = 70$ (E. Clayton, unpublished observations).

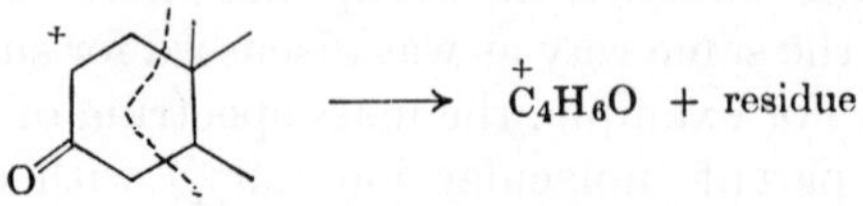

The triterpenes, on the other hand, have been investigated less extensively (T. Bryce, and R. I. Reed, unpublished observations, Budzikiewicz, Wilson and Djerassi, 1963). Here again it is possible to select only one or two examples which are invaluable in analysing the structure. The spectrum of methyl-15-dehydro-oleanolate-3-acetate possesses two abundant ions at $m/e = 260$ and 201. The explanation for these is as follows:

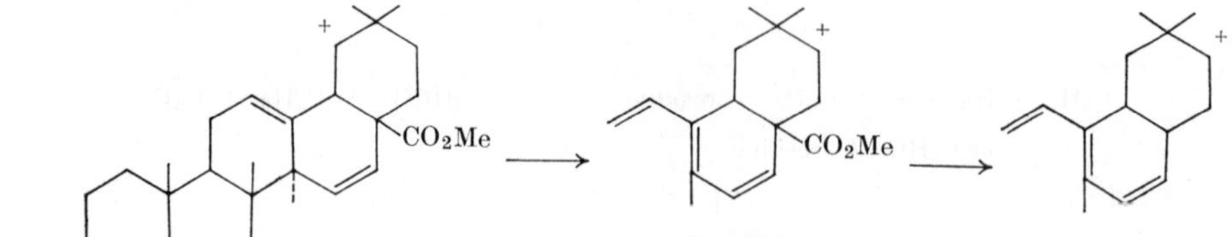

$m/e = 260$ $m/e = 201$

The ion 260^+ is formed by a retro-Diels-Alder that then loses $m/e = 59$, which it will do readily as the carbomethoxyl group is joined at a quaternary centre by an allylic bond, to yield the second ion (201^+), the base peak of the spectrum.

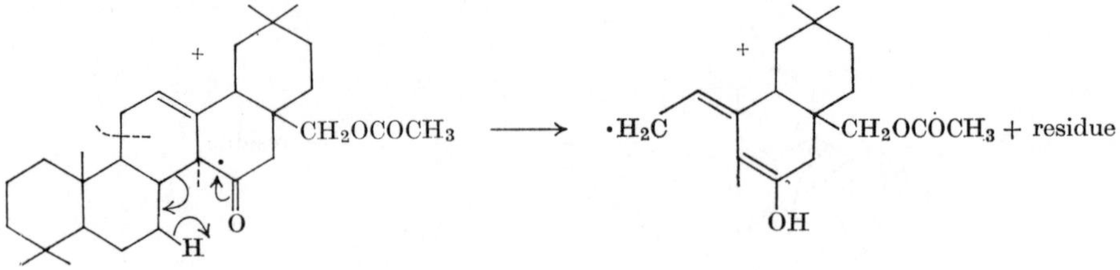

$m/e = 468$

A slightly more complicated example occurs in the fragmentation of 15-ketoerythrodiol diacetate, which shows the very abundant ion $m/e = 291$.

Seemingly, this is not the result of a retro-Diels-Alder, but rather fragmentation of the 9—11 bond which, after yielding an intermediate ion resulting from a hydrogen shift, cleaves.

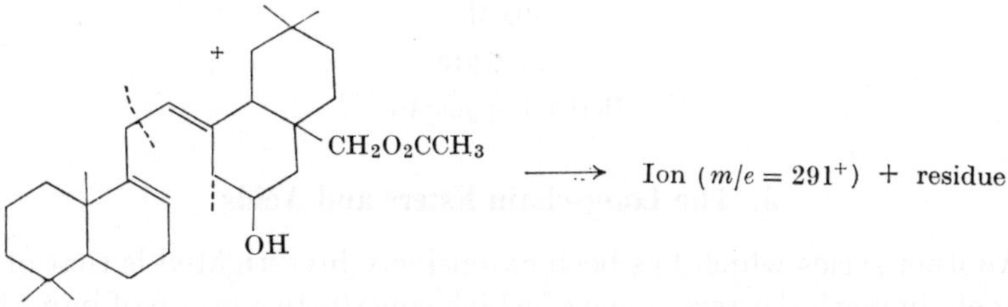

H. The Pimaric Acids

The mass spectra of some diterpene acids, as their methyl esters, have been reported (Bruun, Ryhage, and Stenhagen, 1958; Genge, 1959); two such compounds are methyl pimarate and methyl isopimarate.

The former of these has a molecular ion of low abundance $m/e = 316$ ($C_{21}H_{32}O_2^{+}$) and an abundant ion, the base peak, at $m/e = 121$; two rather moderate ions 181^+ and 180^+ are also present. These ions accord well with the preferred structure of methyl pimarate (Ireland and Schiess, 1963), which is as follows:

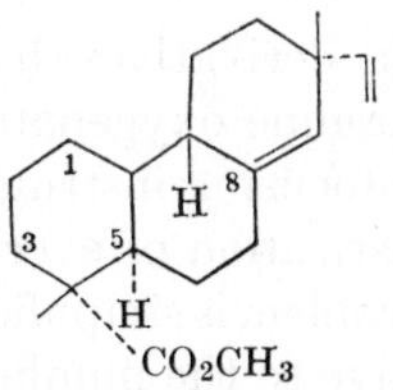

The spectrum of methyl isopimarate differs considerably, much more so in fact than the suggestion that they were merely epimers at carbon-9 would warrant. The parent molecular ion is more abundant, and the base peak occurs at $m/e = 241$; moreover, many ions of lower mass now make a considerable contribution to the spectrum. More recently, the theory that the two systems are simply epimeric has been rejected and a new formula more in line with the observed cracking pattern confirmed

(Antkowiak, Apsimon and Edwards, 1962; Ireland and Newbould, 1962).

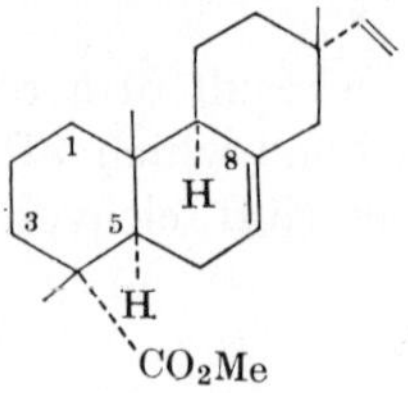

$m/e = 316$

Methyl isopimarate

J. The Long-chain Esters and Acids

Another series which has been extensively investigated is that of the long chain methyl esters, some of which constitute a group of important natural products (Ryhage and Stenhagen, 1963).

Again, in such a voluminous set it is hardly possible to give an adequate idea of all the variations in fragmentation by the examination of a single member. However, some conception of the value of the method may be obtained by its use in the analysis of phthiocerol. The substance, which was known to be a mixture, was separated into constituents and each degraded to a hydrocarbon. Mass spectrometric examination then showed the molecular ions to be the 4-methyl substituted dotria- and tetratriacontanes (Ahlquist *et al.*, 1959; Ryhage, Ställberg-Stenhagen and Stenhagen, 1959a,b).

K. The Inositols and some Sugar Derivatives

The inositols and sugar derivatives have been investigated. In general, it is difficult to examine oxygenated molecules by mass spectrometry since, with single-focusing instruments, it was rarely possible to decide the chemical constitution of every fragment ion. In the two classes here discussed, the problem is simplified because there are so many oxygen atoms present relative to the number of methylene groups that the identification of fragment ions is usually easy. The main difficulty was that the materials were rather involatile, and therefore they had to be run on a probe inserted directly into the ion chamber. This technique favoured the production of ions by secondary processes, viz. ion–molecule collisions, and provided a molecular weight from the $(P+1)^+$ ion even where, as is usually the case, the parent molecular ion was absent.

The probable course of the fragmentation was made easier to follow by the analysis of a series of isomers (Reed, Reid and Wilson, 1962), Thus, in the series of inositols the main fragmentation sequence was deduced by this method (see Appendix 2, p. 193). Comparable investigations have been carried out in the sugar glycosides, in which series the large number of oxygens again aids analysis (Finan, *et al.*, 1964).

L. The Alkaloids

The analysis of nitrogen-containing compounds has been the subject of many systematic investigations in at least three schools, while institutions which do not possess mass spectrometers have availed themselves of the information that fragmentation patterns will provide (Biemann, 1961; Biemann and McCloskey, 1962; Biemann and Spiteller, 1962; Biemann, Burlingame and Stauffacher, 1962; Biemann, De Jongh and Schnoes, 1963; Biemann, Schnoes and McCloskey, 1963; Biemann, Spiteller-Friedmann and Spiteller, 1963; De Jongh and Bieman, 1963; Friedmann-Spiteller and Biemann, 1961; Gorling, Burlingame and Biemann, 1963; Schnoes, Burlingame and Biemann, 1962; Clayton and Reed, 1963, 1964.

Two important functions are fulfilled by mass spectrometric analysis, namely the determination of molecular weight and (usually) the identification of the nucleus of the alkaloid. The main feature of analysis has been to determine the structure of alkaloids, although other types of nitrogen compound have also been examined. Two examples will give some idea of the application of the method.

The mass spectrum of sarpagine gave a rather abundant parent molecular ion $m/e = 310$. This molecule was not, however, the most suitable for investigation, so it was methylated, the hydroxymethyl converted to methyl, and the isolated double bond reduced. The mass spectrum of methoxy dihydrosarpagine shows a parent molecular ion $m/e = 310$ and a base peak at $m/e = 198$ ($C_{13}H_{12}NO^+$). Comparison of the spectrum with that obtained from a degradation product of ajmaline (an alkaloid of known constitution) showed a remarkable similarity. The corresponding ions in the spectra of each derivative were present in similar abundance, with the difference that the fragment ions in the sarpagine derivative were uniformly sixteen mass units higher. The degradation product of ajmaline has the structure, suggesting that the

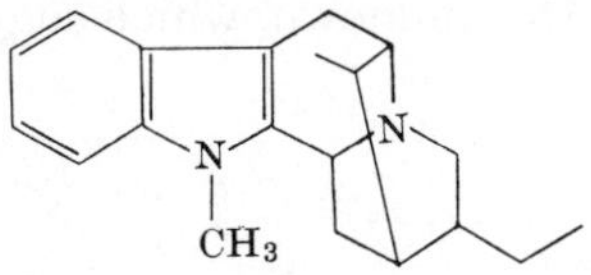

sarpagine derivative has a corresponding arrangement. However, the sarpagine compound is known to possess a methoxyl group absent in the ajmaline product. If allowance is made for this, the sarpagine compound lacks one methylene group compared with the above. The ion $m/e = 182$ in the ajmaline structure may reasonably be assigned to the indole nucleus, which possesses an *N*-methyl that must be absent in sarpagine. On the other hand sarpagine possesses a methoxyl group which is most probably attached to the aromatic ring as it remains with the indole fragment. Therefore the sarpagine derivative is most likely

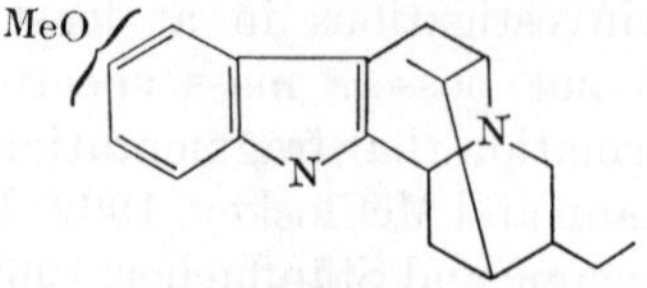

For the detailed argument that positions the methoxyl and the double bond reference should be made to other works (Biemann, 1962a, p. 309).

A further example employing similar methods has been reported in the elucidation of the structure of vincaddiformine. The spectrum is characterized by an abundant molecular ion $m/e = 338$, and an exceedingly prominent ion 124^+ ($C_8H_{14}N^+$) which is the base peak of the spectrum. It was already known that this ion together with $(P-28)^+$ are characteristic of the aspidosperma type alkaloids. It is therefore reasonable to consider the present compound as a member of the series

$m/e = 354$

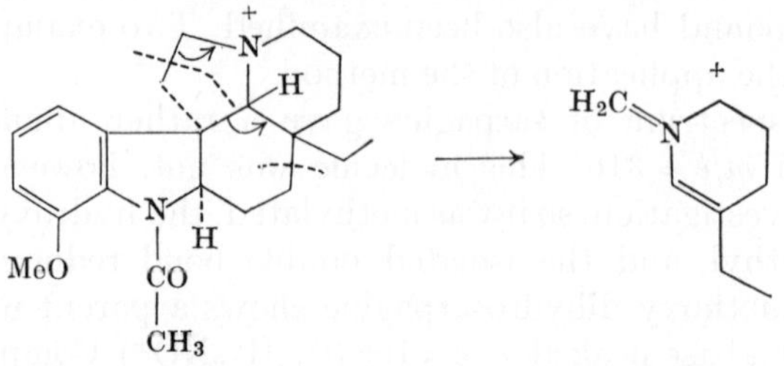

$m/e = 124$

Aspidospermine

and, since the base peak, which is the same in both the known and unknown, has the origin shown above, it follows that vincaddiformine will have the same structure for this part of the molecule. The other ion considered characteristic of such alkaloids $(P-28)^+$ is formed by the elimination of ethylene.

It does not appear in the unknown, which suggests that there is some

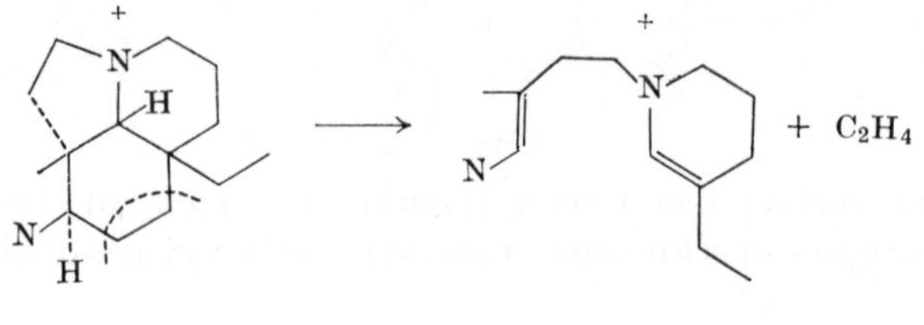

structural variant to produce it. The molecular weight of aspidospermine is 354.

Reduction of the compound under examination to dihydrovincaddiformine permits the corresponding elimination. It is known that one molecule only of hydrogen is taken up, therefore the structural feature which hindered elimination in vincaddiformine was a double bond. Now, it is well known that adjacent double bonds strengthen a carbon–carbon single bond and, therefore, the double bond which prevents the loss of ethylene is either next to the elided group or contained in the group itself.

When the removal of the two-carbon residue is observed in the dihydroderivative, the mass lost is eighty-six rather than twenty-eight units. It follows that the two-carbon fragment carries a further group of fifty-eight mass units. The information does not allow us to decide whether this is one or two substituent groups, or even the probable constitution of these. Such information can be obtained only by additional chemical or physical techniques when, as in the present example, the mass spectrum was obtained upon a single-focusing instrument. A double focusing mass spectrometer would reveal the constitution of the entire piece lost in forming the $(P-86)^+$ ion as $C_4H_6O_2$, from which its structure

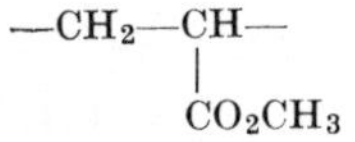

is fairly readily obtained.

In view of the fragmentation pattern all modifications to the aspidospermine skeleton, at least as far as the right-hand side of the molecule is concerned, are now accounted for. Comparison of the molecular weights, having made allowance for the double bond and substituent, reveals that as regards the remainder of the molecule, aspidospermine has a mass thirty units greater than vincaddiformine. Therefore, the latter molecule does not contain the methoxyl group known to be in the former. Accordingly the partial structure is of the form

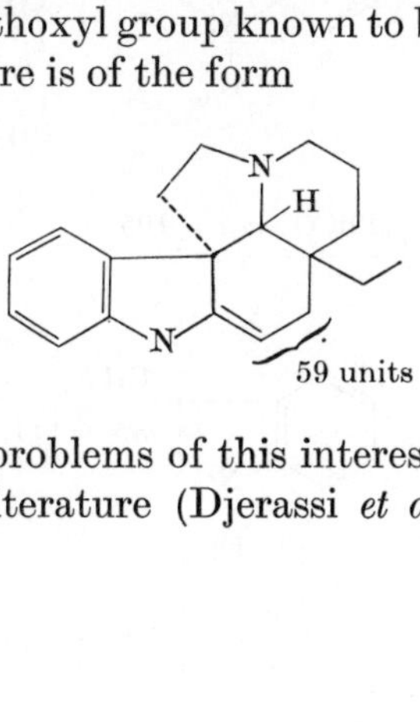

A further study of the problems of this interesting class of compounds will be found in the literature (Djerassi *et al.*, 1962f).

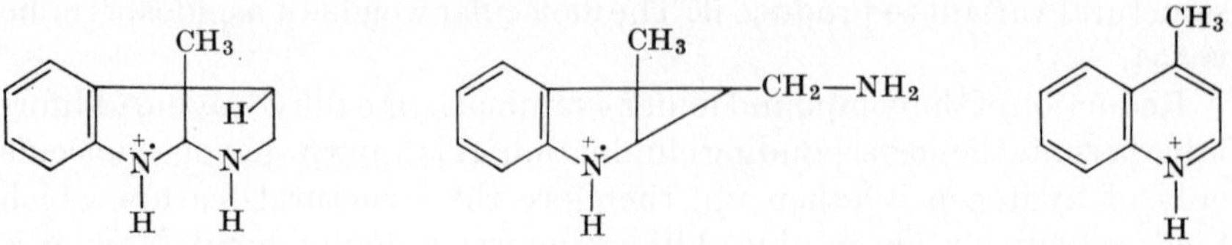

Desoxybisnoreseroline

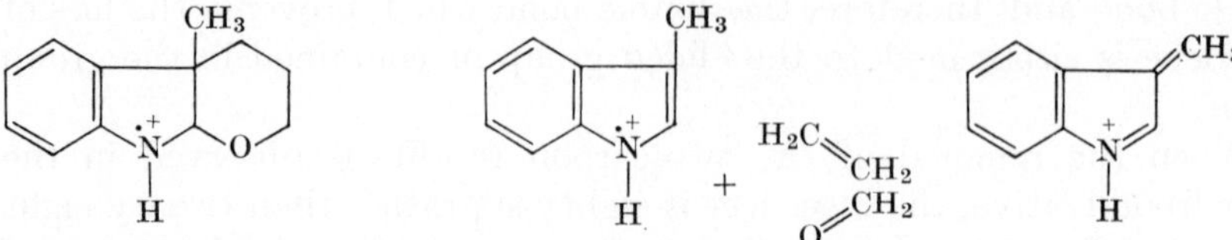

3-Methylhexahydropyrano-
[2,3b] indole

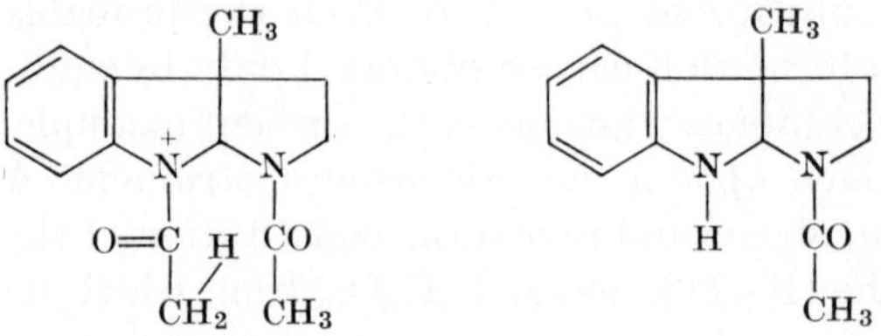

N,*N*-Diacetyldesoxybisnoreseroline

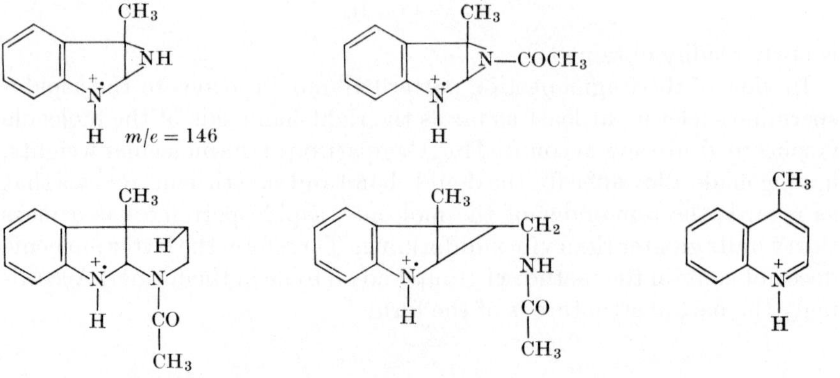

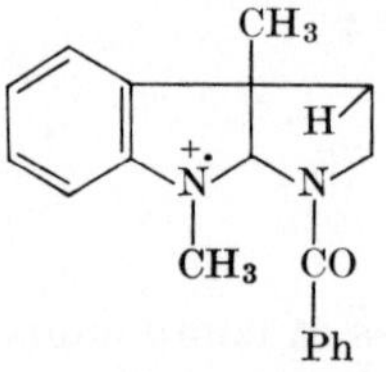

N(b)-Benzoyldesoxy
noreseroline

PhCO$^+$ $m/e = 105$

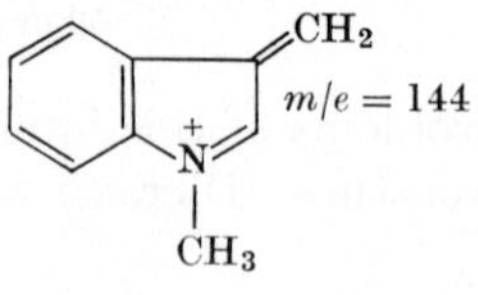

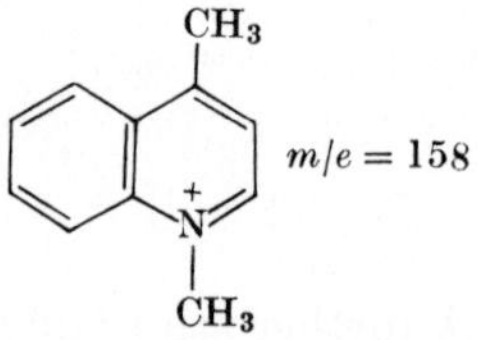

The application of mass spectrometry to the heterocyclic systems has also found important applications in the chemistry of natural products by an examination of the cracking patterns of known structures, which enables fragmentation pattern and structure to be correlated.

Some correlation studies have been carried out upon a whole series of pyrrolo–indoline systems (Clayton and Reed, 1963).

In all cases it may be plausibly deduced that fission at a multiply substituted carbon occurs to the greatest relief of strain. This is certainly supported by the observation that, whereas the parent molecular ions of desoxybisnoreseroline and of desoxynoreseroline are also the base peaks of the spectra (Clayton and Reed, 1963), the abundance of the ion in the acetylated and benzoylated derivatives is much less. The addition of a side chain as in physostigmine further lowers the abundance. Comparison with an even more branched compound, e.g. eserethole methine, results in a still less abundant molecular ion.

Fragmentation with re-arrangement occurs in the break-down of diacetyldesoxybisnoreseroline to yield a molecular ion identical to *N*(b)-acetyldesoxybisnoreseroline, which then cleaves with further re-arrangement to produce the ion $m/e = 146$.

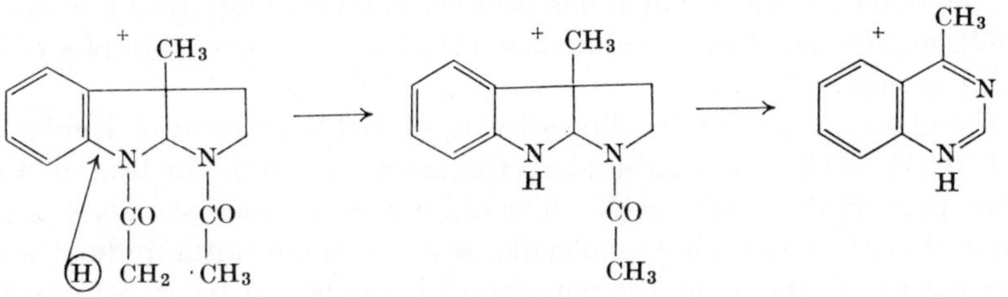

$m/e = 258$ $m/e = 216$ $m/e = 146$

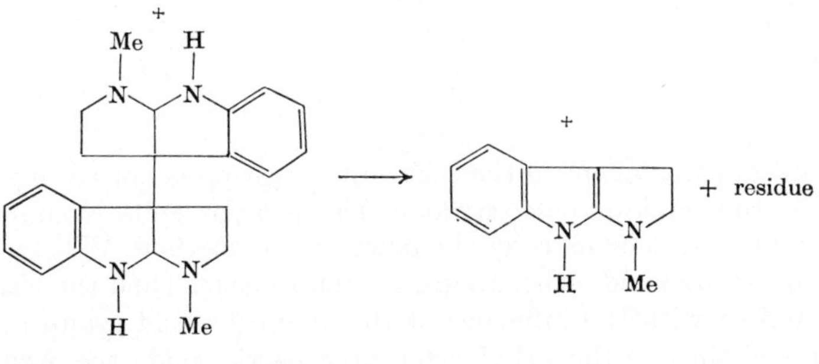

This sequence, which is based upon the thermodynamic argument of stability of the ions and neutral fragments formed, is consistent rather than compelling. It does, however, offer a guide to the fragmentation patterns arising from similarly constituted structures. Accordingly, the fragmentation of chimonanthine may be easily understood.

M. Amino Acids and Peptides

A mass spectroscopic examination of the amino acids is particularly valuable as most other methods of spectroscopic examination meet considerable difficulties. Unless there is a chromophore attached to the amino acid, ultraviolet spectroscopy is of no assistance and the infrared spectrum often encounters serious problems. The application of mass spectroscopic methods is not easy, for the free acids exist as dipolar ions and as such have very low vapour pressures. Two methods are then available. One is to pyrolyse the amino acid or peptide and examine the diketopiperazines so produced; the second is to increase the volatility of the free acid in some way. The former method has been reported (Svec, 1965; R. M. Sathe, unpublished observations) and is being further exploited in the author's group; a double-focusing instrument is being used to determine the chemical composition of the ions formed.

The second method which has been extensively employed (Biemann, 1962), makes use of the α-amino acid ethyl esters. Some examples of the latter are given below.

The mass spectrum of proline ethyl ester yields a parent molecular ion 143^+ ($C_7H_{13}NO_2^+$) and an abundant fragment ion (70^+). The fragment ion is the pyrrolinium ion $C_4H_8N^+$. The difference between the mass of this ion and that of the whole molecule, seventy-three units, indicates the formula $CO_2C_2H_5$, since the compound is known to be an ethyl ester. Therefore the structure is

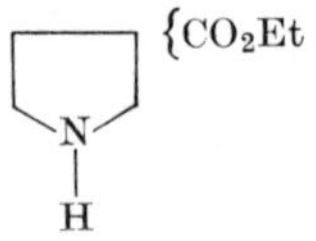

The point of attachment of the carboethoxy group is not readily determined by the cracking pattern alone. This is a particularly simple case which, however, demonstrates the power of the method. When a longer side chain is involved, other fragmentations occur. Then the cleavages are in accord with the influence of the amino or acid group upon an aliphatic chain. In the ethyl ester of aspartic acid, the prominent ion $m/e = 88$ arises by the elimination of ethylene from the base peak of the spectrum.

The first fission is consistent with the presence of a secondary amine and the second with the presence of a carboxylic ester grouping.

$$H_5C_2OCOC\overset{+}{H}_2CH(NH_2){\vdots}COOC_2H_5 \quad m/e = 189 \longrightarrow H_5C_2OCO\overset{+}{C}H_2CHNH_2 + \text{residue} \quad m/e = 116$$

$$H_5C_2OCO\overset{+}{C}H_2CHNH_2 \quad m/e = 116 \longrightarrow C_2H_4 + HO_2C\overset{+}{C}H_2CHNH_2 \quad m/e = 88$$

This type of analysis may be extended to include sulphur-containing amino acids. Methionine ethyl ester gives a very abundant ion $m/e = 61$ corresponding to the well established fission β to the sulphur atom. Ions are also formed by fission β to the amino group.

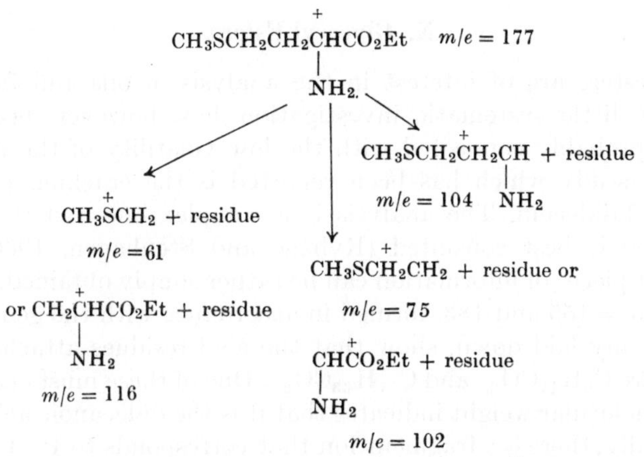

Some of the fragment ions so formed decompose still further and build up the rather complicated cracking pattern observed.

When the problem is applied to peptides, the analysis necessarily becomes more involved, for it is now necessary not only to identify the individual amino acids, but also to determine their sequence. Two methods have been reported (Biemann, 1962a, p. 283). One is limited to the examination of the spectra of small peptides which contain up to three units. This requires the conversion of the free acid to an ester (the ethyl ester is commonly employed) and the formation of a trifluoroacetyl derivative of the free amino groups. The peptide is now sufficiently volatile for mass spectrometric analysis.

The other method is to reduce the peptide to the polyamino alcohol by means of lithium aluminium hydride. The cracking pattern of the alcohol or its acetyl derivative is then analysed. These compounds undergo cleavage in the positions one might expect,

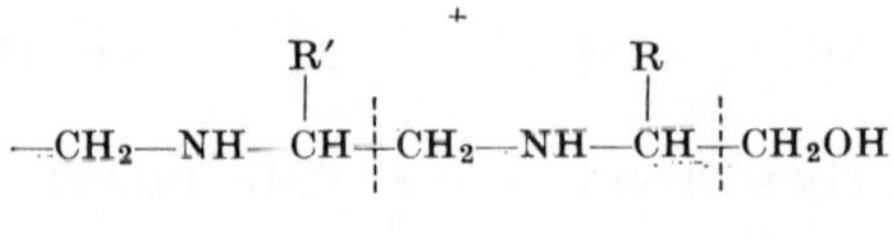

namely β to the nitrogen. One finds in the mass spectrum ions corresponding to these fragments or to their loss from the parent molecular ion. In addition, other fragmentations occur which result from the elimination of the groups R, R′ etc; re-arrangements have also been reported (Biemann, Gapp and Seibl, 1959; Biemann and Vetter, 1960).

N. Glyceryl Esters

These esters are of interest in the analysis of oils and fats. Comparatively little systematic investigation has, however, been made, which is probably connected with the low volatility of the materials. One such study which has been reported is the cracking pattern of 2-lauro-1,3-didecoin. The analysis is a complex one and the original publication is best consulted (Ryhage and Stenhagen, 1960). Three important pieces of information can be rather simply obtained. The acyl ions at $m/e = 155$ and 183, formed in accordance with the general principles already laid down, show that the acid residues attached to the glycerol are $C_9H_{19}CO_2$– and $C_{11}H_{23}CO_2$–. One of these must occur twice, and the molecular weight indicates that it is the didecanoic acid derivative. Thirdly, there is a fragment ion that corresponds to $P—CH_2O_2C—C_9H_{19}^+$, but none corresponding to an analogous structure involving lauric acid. The position of the three acids in the structure is thus confirmed, the $-CH_2—O_2CC_9H_{19}$ being formed by cleavage in the glyceryl part of the molecule.

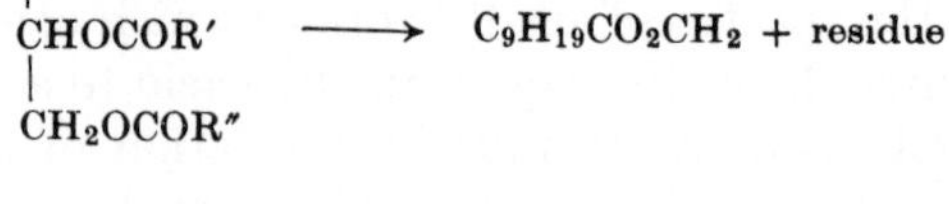

where either R = R′, or R = R″, which places the decanoic acids at the 1,3-positions.

$$\begin{array}{l} C_9H_{19}CO_2CH_2 \\ \qquad\quad | \\ \qquad\quad CH.O_2C.C_{11}H_{23} \\ \qquad\quad | \\ C_9H_{19}CO_2CH_2 \end{array}$$

The structure of the alkyl chains in each acid would require a very detailed examination of the cracking pattern if indeed this would suffice.

P. Lupulone

The spectrum shows a molecular weight of 414 and the fragment pattern consists principally of the loss of alkyl chains. The base peak occurs at $m/e = 69$ ($C_5H_9^+$), and there is a corresponding ion at $(P-69)^+$. Other fragment ions in decreasing order of abundance occur at 41^+, $(P-85)^+$, $(P-127)^+$, $(P-139)^+$ and $(P-141)^+$. The ions at 69^+ and at $(P-69)^+$ represent the loss of a pentenyl chain, which presumably joins at a branch in the carbon skeleton. The ion 41^+ ($C_3H_5^+$) is unlikely to be a further branch substituent as there is no ion $(P-41)^+$ corresponding to loss of this. Three carbon fragments are known to cleave preferentially even in straight-chain alkanes. The ion $(P-85)^+$ refers to the loss of an acyl group (since the overall structure of lupulone contains no less than four oxygen atoms). Double-focusing instruments would reveal the presence of an oxygen in the eighty-five units elided. The absence of ions corresponding to 127^+, 139^+, and 141^+ suggests that these fragments are not eliminated as single units, but compounded of smaller molecules. The use of a double-focusing mass spectrometer would aid somewhat, but it is doubtful if it would allow a structure determination without some additional information.

The formula of lupulone is

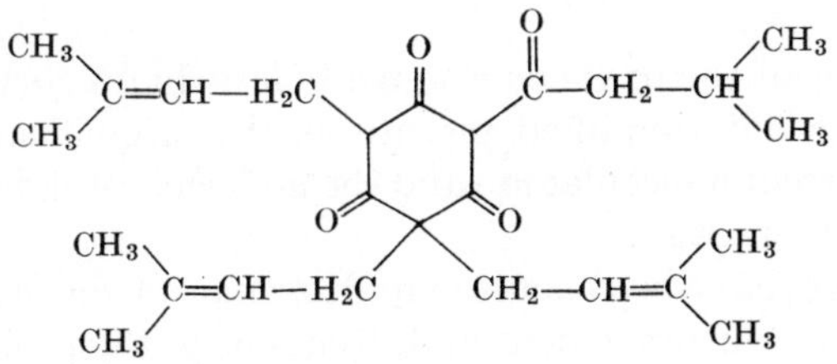

Q. Reduction

There are perhaps two techniques which, above all others, are useful aids to the mass spectrometrist in the elucidation of structure. One of these, which has been but little used, follows the classical chemical method of removing all the substituents and determining the nature of the carbon skeleton. In the conventional method, the material was treated rather drastically (usually distilled with sulphur or selenium) in order to produce a substituted aromatic compound which then became the basis for the reconstruction of the original molecule. In principle, the same technique may be followed when the structural determination relies on mass spectrometry. In this instance, the production of the carbon skeleton need not be achieved by such drastic methods since, as already stated, it is easier to determine the structure of a cycloalkane than of an aromatic polycyclic hydrocarbon. The method seems to have been used in only one large-scale operation so far; which is in the elucidation of the structure of fungichromene (Cope *et al.*, 1962). All oxygen functions were removed and the double bonds saturated by hydrogenation. The spectrum obtained was that of a dimethyl triatriacontane. Two abnormally abundant ions, having regard to their position in the spectrum, namely $m/e = 197$ ($C_{14}H_{29}^+$) and 323 ($C_{23}H_{47}^+$), together with the re-arrangement ion at 196^+ indicate the position of one of the substituent methyl groups

$$C_{21}H_{43}\text{—}\overset{+}{C}H(CH_3)\text{—}C_{12}H_{25}$$

Similarly abundant ions at $m/e = 113$ ($C_8H_{17}^+$), and 417 ($C_{29}H_{59}^+$) locate the other. The full structure of the carbon skeleton was

$$H_{13}C_6\text{—}CH(CH_3)\text{—}(CH_2)_{13}\text{—}\overset{+}{C}H(CH_3)\text{—}C_{12}H_{25}$$

The assignment of the oxygens was made largely on chemical grounds. This is a somewhat simplified picture as the original alkane gave an insignificant parent molecular ion and the molecular weight was obtained in a rather different way.

Other uses too, have been made of hydrogenation. Saturated molecules are less liable to re-arrangement and, frequently, the position of hetero-atoms within the skeleton may be more easily determined.

One example is the comparison of the cracking patterns of furfuryl and tetrahydrofurfuryl alcohol. The former gives rise to fragments by a re-arrangement process that makes it difficult to locate the original position of the hydroxymethylene with respect to the oxygen.

TABLE XXIII

m/e	α-Furfuryl alcohol	α-Tetrahydro-furfuryl alcohol	m/e	α-Furfuryl alcohol	α-Tetrahydro-furfuryl alcohol
18	—	16·2	51	23·1	—
27	48·0	43·1	52	17·8	—
29	59·5	35·3	53	65·9	—
31	38·6	31·31	69	33·8	—
37	20·6	—	70	34·4	—
38	32·2	—	71	—	100
39	96·8	30·2	81	15·7	—
40	12·0	—	82	13·5	—
41	97·8	54·2	97	55·1	—
42	77·3	30·3	98	100	—
43	21·5	77·4	102	—	0·21
44	12·8	13·2			

Knowing the structure, it is of course possible to produce a plausible mechanism for the second most abundant ion in the spectrum,

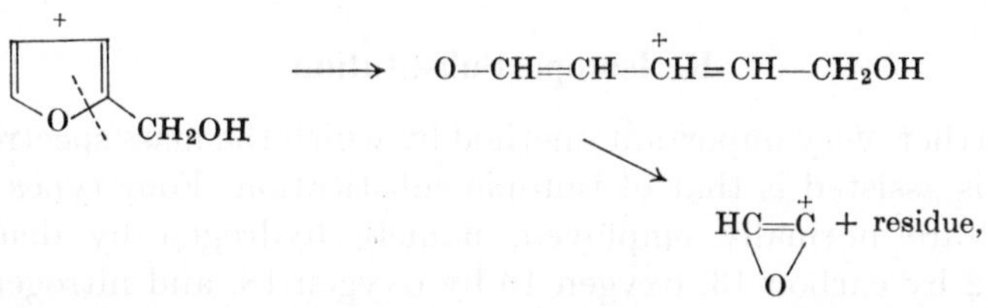

but it is by no means as convincing as the cleavage observed in the tetrahydro derivative for, in the reduced compound, no such re-arrangement difficulty exists. The known factors already listed indicate that the hydroxymethylene group will cleave readily if attached to a ternary or quaternary atom. This is confirmed by the base peak $m/e = 71$ which corresponds to the fragmentation

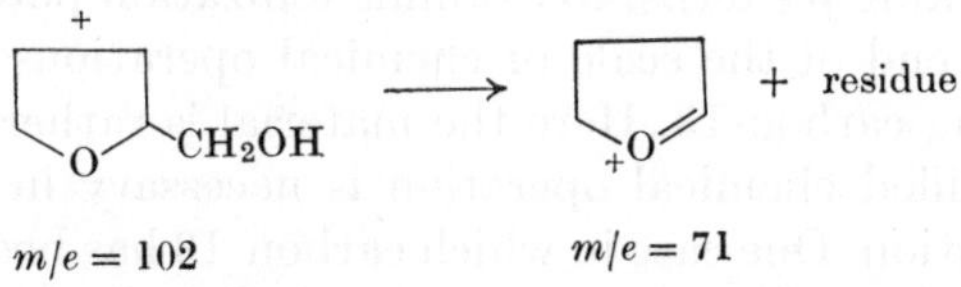

Moreover, the presence of the ion $m/e = 41$, corresponding to the loss of sixty-one mass units, requires there to be three unsubstituted methylene groups in a chain, a condition met only in α-tetrahydrofurfuryl alcohol.

The mode of decomposition is as follows:

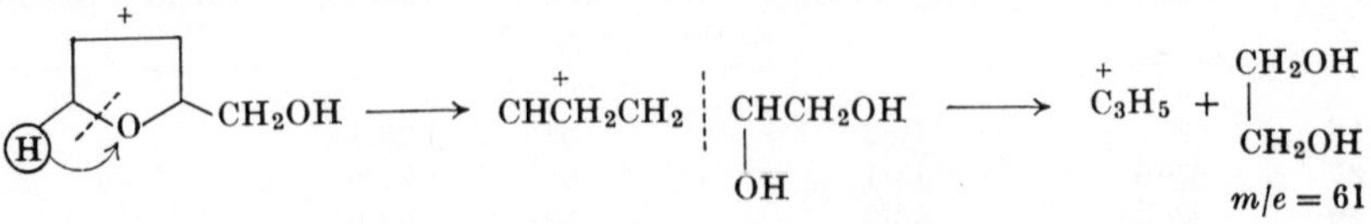

a type of fission which is frequently encountered in the cracking patterns of cyclic oxide rings. The alternative ring breaking occurs also.

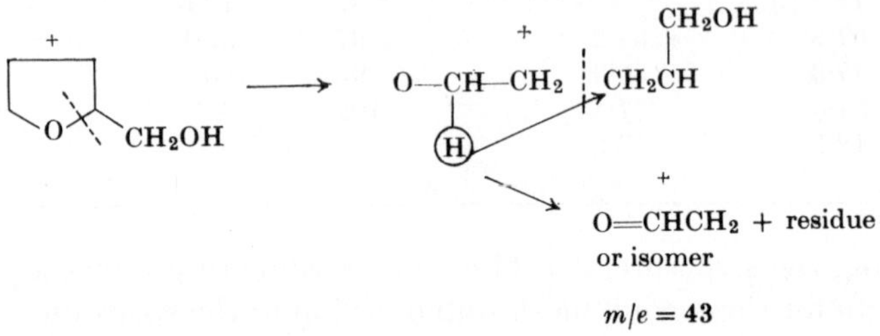

R. Isotopic Substitution

The further, very important, method by which the mass spectrometric analysis is assisted is that of isotopic substitution. Four types of substitution are normally employed, namely hydrogen by deuterium, carbon-12 by carbon-13, oxygen-16 by oxygen-18, and nitrogen-14 by nitrogen-15. Of these possible methods, the first is by far the most popular. By and large it is the easiest chemical change to bring about. Deuterium is freely available, reasonably priced, and there exist relatively simple chemical operations to introduce it into the molecule; the methods (exchange, hydrogenation, and so on) are not difficult to carry out on milligram quantities of material. Consequently, deuteration has been employed for a wide variety of uses: to determine structure, to investigate cracking patterns, to examine ionization potentials, etc.

At the other end of the scale of chemical operations comes isotopic studies involving carbon-13. Here the material is rather expensive and considerable skilled chemical operation is necessary in order to carry out the substitution. One case in which carbon-13 has been incorporated is [2-^{13}C]phenylglyoxal.

It is well known that phenylglyoxal re-arranges to the mandelate anion under the influence of basic catalysts. A mass spectrometric analysis of carbon-13 enriched compounds was used to investigate the mechanism of this internal Cannizzaro reaction (Doering, Taylor and Schoenewaldt, 1948). The carbon-13 labelled phenylglyoxal was converted to the mandelate ion and thence to mandelic acid, which was further oxidized to benzoic acid. A mass spectrometric analysis proved that the benzoic acid was still enriched, and the carbon dioxide produced by the oxidation had the normal carbon-13 abundance.

$$C_6H_5{}^{13}COCHO \xrightarrow{\text{Base}} C_6H_5{}^{13}\underset{H}{C}(OH)CO_2^-$$

$$C_6H_5{}^{13}CH(OH)CO_2H \xrightarrow{\text{Oxidation}} C_6H_5{}^{13}CO_2H + CO_2$$

This demonstrated that there was no re-arrangement of the carbon skeleton on hydroxylation. Similar studies were carried out using deuterium to show that the hydrogen shift was intramolecular, and the details of these are in the original publication which should be consulted (Doering, Taylor and Schoenewaldt, 1948).

The oxygen-18 replacement reaction has been little used. Most, if not all, the published works have been occupied with questions of mechanism of ester hydrolysis (Bender, 1951), oxygen take up in suitably disposed hydroxyketones and in lactone formation (Dahn, Moll and Menassé, 1959; Bender *et al.*, 1961).

The use of nitrogen-15 equally has been concerned with reaction mechanisms such as nitrogen fixation in plants (Dugdale and Neess, 1961) or the study of re-arrangements occurring at a nitrogen atom (Prosser and Eliel, 1957). There have also been some interesting structure determinations that were greatly assisted by the use of nitrogen-15 enriched molecules.

The known decomposition of phenylhydrazine when distilled at atmospheric pressure in the absence of nitrogen has been examined by this means. A mass spectrometric examination of the products showed that the reaction followed the path

$$2C_6H_5{}^{14}NH^{15}NH_2 \rightarrow C_6H_5{}^{14}NH_2 + {}^{15}NH_3 + {}^{14}N^{15}N + C_6H_6$$

The mechanism which is consistent with this isotopic distribution requires that one molecule of phenylhydrazine loses nitrogen, reduces the phenyl group to benzene and also reduces a second molecule of phenylhydrazine to aniline and ammonia.

$$C_6H_5{}^{14}NH^{15}NH_2 \rightarrow C_6H_6 + {}^{14}N^{15}N + 2H$$

$$C_6H_5{}^{14}NH^{15}NH_2 + 2H \rightarrow C_6H_5{}^{14}NH_2 + {}^{15}NH_3$$

S. Deuteration

The very simplest reaction, and still one of great practical use, is the exchange reaction of deuterium for hydrogen.

Perhaps the most obvious application of this method is the determination of the position of a keto-group in a cyclic ketone. As is well known, the hydrogens upon neighbouring carbons are activated by the presence of the ketone function and will undergo a base catalysed exchange. The mass spectrum of the material is obtained, the exchange is carried out exhaustively, and the mass spectrum redetermined. If the molecular weight has increased by four mass units, four hydrogens have been exchanged and therefore the ketone is present in a structure of the form $-CH_2-CO-CH_2-$. Similarly, the environment may be determined for other exchanges. The peculiar advantage of this approach is that the deuterium alkoxide used for the exchange need not be pure, nor need the exchange itself be quantitative. The presence of any contribution to each of the possible four reaction sites will be recorded by the mass-spectrometer as a molecular ion four units greater than that originally found. Should there be no parent molecular ion, the problem is not so easy. Nevertheless, provided that the highest ion recorded (excluding isotope contributions) upon the original compound contains the keto function (double-focusing instruments make a valuable contribution here), any exchange will be observed when the spectrum of the deutero sample is compared with that originally obtained. Should the mass of the ion have increased by four units the $-CH_2COCH_2$ system must again be present. However, if only two atoms have been incorporated, then two or more possibilities obviously exist: the original structure is of the form $-CH_2-CO-C=$ or $=CHCOCH=$; or one is dealing with a fragment ion

$$-CH_2\overset{+}{C}O \,\vdots\, \overset{|}{\underset{|}{C}}H$$

the result of fission α to the keto-group. Both conclusions may be correct, but a final decision would need very careful analysis.

In a similar way, deuteration may be immensely helpful in elucidating the cracking pattern and hence the structure of nitrogen containing compounds. A primary or secondary amine function can be readily picked out of a structure in this way since the hydrogen attached will quickly equilibrate with deuterium. Now comparison of the spectra permits identification of the fragments containing nitrogen which, having the hydrogen replaced by deuterium, will be one or two mass units greater. This technique is particularly useful if the original molecule contains two nitrogen atoms. Although double-focusing instruments can reveal the presence and number of nitrogen atoms in fragment ions, when only one of a pair of nitrogens is present in a fragment this technique alone cannot solve the problem. A trivial example which nevertheless shows the method of application occurs in the cracking pattern of nicotine (see Chapter 4, p. 94).

TABLE XXIV

Mass spectrum of nicotine

m/e	Relative intensity	*m/e*	Relative intensity	*m/e*	Relative intensity
26	2·30	56	1·08	84($C_5H_{10}N^+$)	100·00
27	6·43	57	0·72	89	0·67
28($CH_2N_9C_2H_4^+$)	17·05	58	0·38	90	0·60
29	1·68	59	0·48	91	1·94
30	1·34	62	0·94	92	4·99
31	0·24	63	3·36	93	0·96
32	2·30	64	1·63	104	1·56
36	0·38	65	4·89	105	1·51
37	0·73	66	1·03	106	0·74
38	2·30	67	1·08	107	0·60
39	8·99	68	0·58	116	0·48
40	1·68	69	0·62	117	2·66
41	5·68	70	0·24	118	3·31
42($C_2H_4N^+$)	20·12	74	0·24	119	5·16
43	1·85	75	0·43	130	3·21
44	2·76	76	0·67	133($C_8H_9N_2^+$)	26·59
49	0·24	77	2·28	144	0·38
50	3·00	78	3·53	145	0·34
51($C_4H_3^+$)	7·46	79	1·37	147	0·43
52	3·41	80	0·84	159	1·20
53	1·49	81	0·91	161	17·63
54	1·51	82	3·55	162($C_{10}H_{14}N_2^+$)	19·14
55	4·29	83	0·36		

A further use of deuteration in determining the origin of particular fragment ions may also be mentioned. It is known that there is a rather abundant ion $(P-43)^+$ in the methyl esters of long-chain fatty acids. The origin is uncertain, but several possibilities exist which may be deduced from correlation studies with known simpler structures. Progressive deuteration on the carbon atoms in these esters (Dinh-Nguyen *et al.*, 1961) has yielded the surprising result that the elimination occurs within the carbon chain. The first, second and third carbon atoms in the alkyl chain are elided, with a single hydrogen re-arrangement

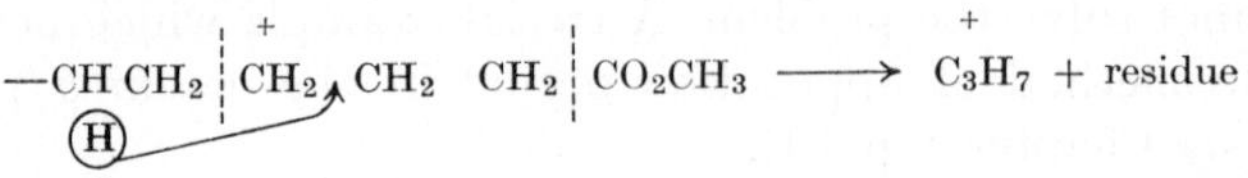

This experiment shows *inter alia* the danger of an incautious use of correlation methods in the interpretation of cracking patterns for, knowing the ion 43^+ to be a hydrocarbon, one could still not infer its origin in the alkyl chain.

Another important use for deuterium labelling would be in the reduction of double bonds. It has already been emphasized on several occasions that the structure of an alkane is often more easily recognized than the corresponding alkene. Reduction to the alkane therefore offers the best means of determining the basic structure. The drawback here, however, is that, by reducing the double bond, the chance has been lost of locating it. If at least some of the original alkene were reduced with deuterium rather than hydrogen, a comparison of the two spectra should locate the position of the bond. The procedure is not quite as easy as it might appear, for it is very important that no re-arrangement should occur on reduction, and that a concomitant exchange of hydrogen and deuterium does not occur on an adjacent carbon (or some other point) in the molecular structure.

T. Hydrogen Migration Reactions

Finally, there has begun recently a systematic study of hydrogen re-arrangement reactions. It has been mentioned, in the many instances of the cleavage of substituted alkanes, that the fragmentation is accompanied by re-arrangement of one hydrogen. Moreover, many fissions in the breakdown of steroids and terpenes are accompanied by single hydrogen migrations. Little formal evidence of the origin of the hydrogen has been provided, although the consensus of opinion has favoured the fissions shown. Now, some systematic studies have been made upon such migrations in amides, amines (Pelah *et al.*, 1963), steroids (Williams *et al.*, 1963), and triterpenes (Budzikiewicz, Wilson and Djerassi, 1963). Some examples are given here. For *N*-butylacetamide, use of deuterium labelling to prepare *N*-butyl-[D_3]-acetamide and a subsequent comparison of the cracking patterns indicates that there are two main routes to the formation of the ion $m/e = 30$ in the spectrum of the unenriched amide. These are

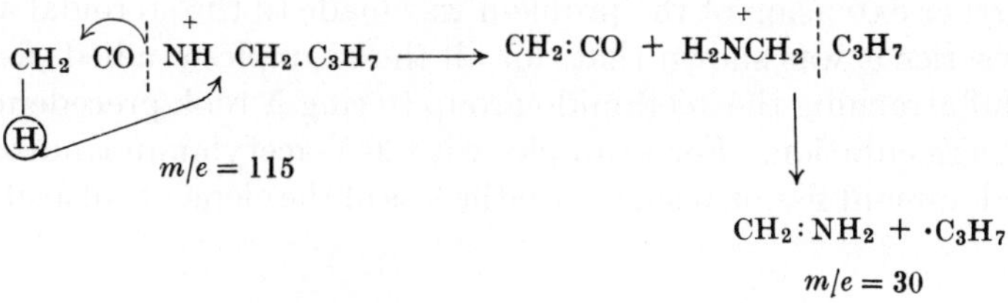

formed by successive cleavages, and

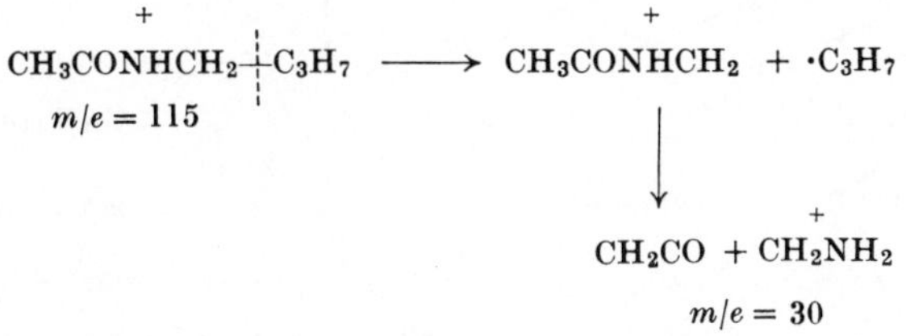

In the deuterated derivative, the ion (30^+) moves almost entirely to $m/e = 31$, showing that the hydrogen in the re-arrangement process derives from the methyl of the acetyl group.

It will be apparent from this study that cleavage also occurs in the *N*-alkyl substituent. Consequently, if it is varied by replacement of the alkyl by a cycloalkyl group then, in order to obtain the corresponding reaction, a further bond must be broken. The same authors have carried out an extensive investigation of the selective replacement of certain hydrogen atoms in *N*-acetylcyclohexylamine. For the full analysis, reference must be made to the original article (Pelah *et al.*, 1963), but it is clear that the need to fragment two bonds substantially alters the fragmentation pattern. The most obvious consequence of the introduction of a cyclo-alkyl group is that the abundant ion now occurs at $m/e = 57$ rather than 31. The result of the isotope substitutions proves that the mechanism of formation is almost certainly

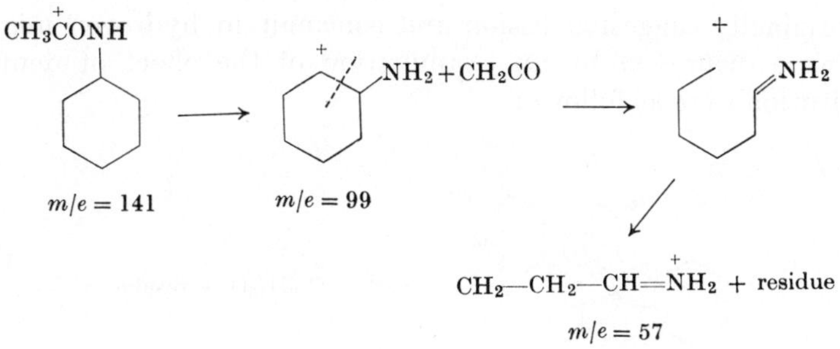

A further extension of the problem was made to the steroidal amines. In this series it was shown that, for all the examples studied, fission of the bond attaching the acetamido group to ring A took precedence over other fragmentations. For example, with 3-*N*-acetylaminoandrostanes, the predominant fission occurs with the loss of the elements of acetamide.

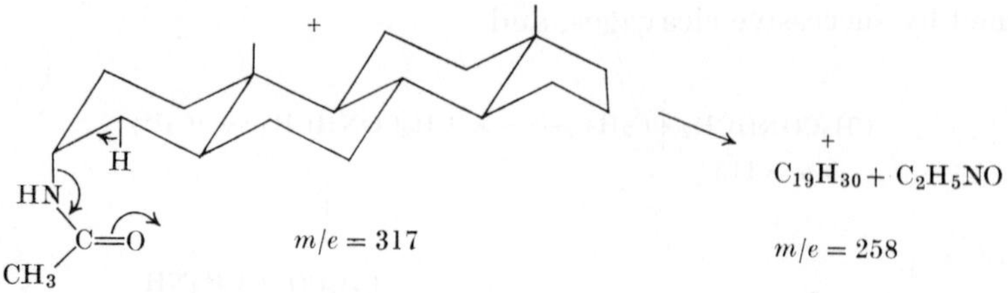

The conclusion to be drawn from this work is that α-bond fission with respect to the nitrogen is very facile.

Similar, extensive studies have been carried out in an analysis of 5α-androstan-11-one where some fifteen different deuterated compounds have been examined (Williams *et al.*, 1963). The conclusions indicate that many of the fragmentation processes are rather complex and diverse. One such analysis is concerned with the elimination of rings A and B from the compound. Cleavage has been shown to occur as follows:

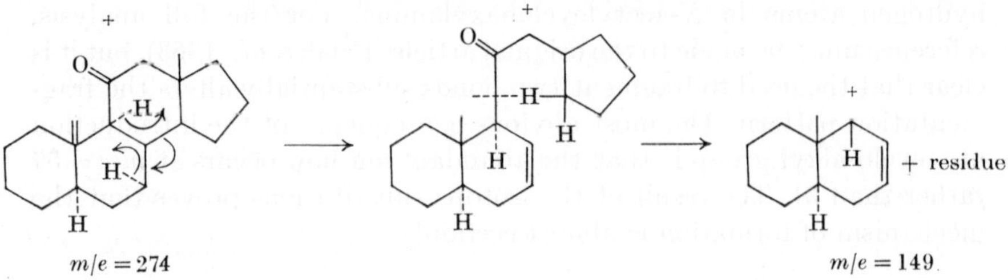

The originally suggested fission and concomitant hydrogen migration that were disproved by an examination of the effect of deuterium substitution were as follows:

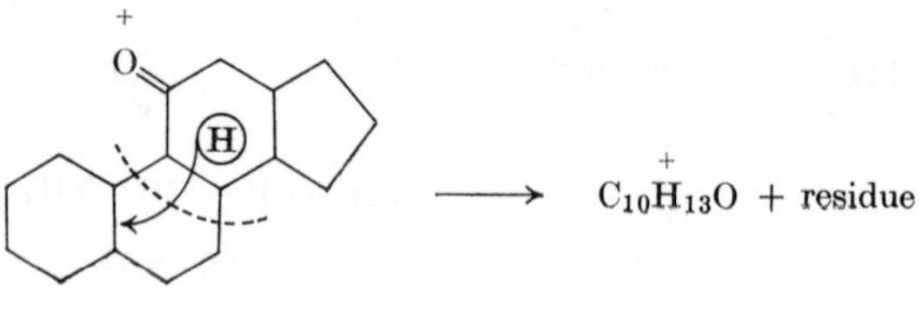

Substitution, on carbon-9 to give 9α-[D_1]-5α-androstan-11-one, at C(12) to give either 12α-[D_1]- or 12β-[D_1]-5α-androstan-11-one, showed that the ion $m/e = 149$ ($C_{11}H_{17}^{+}$) is largely unaffected by these substitutions. Moreover, the spectrum of 8β-[D_1]-5α-androstan-11-one proves that the deuterium is not lost from the ion, since its mass now becomes $m/e = 150$ ($C_{11}H_{16}D^{+}$). These observations are consistent with the new mechanism proposed; other ions are examined in a similar way.

One interesting sidelight upon the enolization of these 11-keto steroids has also been given. The hydrogen abstraction from positions 9α, 12α and 12β has been examined in the course of the synthetic work. It was found that, arranged in a decreasing order of ease of abstraction, $9\alpha > 12\alpha \gg 12\beta$.

CHAPTER 6

The Mass Spectrometry of Mixtures

A. General Considerations

The analysis of mixtures of substances by mass spectrometric methods has had a chequered history. Its importance as an analytical method was early recognized (Washburn and Hoover, 1940), and it has been of particular value to the petroleum industry as a means of estimating the amounts of various hydrocarbons in natural gases and volatile oils. Around 1940, such analyses by chemical methods were indeed a formidable undertaking.

The mass spectrometric method depends upon the following considerations. (i) The mass spectrometer must be stable. The cracking pattern characteristics of any particular molecule run on different occasions under the same operating conditions must be the same or very nearly so. (ii) Linear superposition of the mass spectra of various components which are present in the mixture is necessary. (iii) The sensitivity (s) for the peaks used in analysis must be proportional to the partial pressure (p) of the substance giving rise to the ion. (iv) Discrimination must not be caused by the leak through which the gases pass into the ionization chamber. These requirements are conveniently examined in this order.

It is clear that if condition (i) is not met, then continuous calibration of the instrument will be required, a time-consuming process. Not only would the cracking pattern of every suspected molecule need re-examination, but extensive re-calculations in the mathematical analysis would be necessary.

Some conditions which may affect the cracking pattern include the temperature of the filament and the condition of its surface. The ion repeller voltage is relatively unimportant, particularly in sector instruments where the voltage is kept constant during an analysis. In most instruments, the electron emission from the filament is temperature limited. The admission of a sample to the ion chamber will alter the work function of the filament; this in turn will alter the electron emission, which will lead to a change in the source temperature. Some compounds possess ions, even parent ions, which are temperature dependent, and the cracking pattern is thus changed. The ions of isobutane have been shown to vary with temperature, and the change in the abundance of the parent molecular ion of 2,2,4-trimethylpentane is even more marked (Berry, 1949).

The spatial distribution of electrons from off the filament has been found to affect the cracking pattern. This difficulty is overcome by coating the filament with a stable carbide layer. The filament is "burned in" by running in a hydrocarbon, usually n-butane, until the carbide layer has formed. (Blears, unpublished observations; Mitchell, 1950).

A further problem which is easily overcome by a periodic cleaning of the ion chamber is the deposition of insulating layers upon the inside of the source. The presence of these gives rise to spurious potentials which alter the gradients in the source and hence affect the cracking pattern.

All these variations affect the sensitivity of the peaks. For one compound the sensitivity of all the peaks is the same and so only one measurement upon a suitable fragment ion is necessary in order to determine them all. Absolute measurements are not usually made and it is customary to refer the sensitivities to n-butane as a standard.

Experimenters who have done very large numbers of mixture analyses report that, with careful operation, the short term variations in the cracking pattern (from day to day) are from 2 to 3%; weekly variations of some 5% or more have been observed.

The principle of superposition requires the contributions of each component to be additive in the mass spectrum. The positive ion distribution of one component must not be affected by that arising from others present. The further implication is that the gases do not interfere with each other in transfer from the reservoir to the ion source. These conditions are never realized in practice; somewhere in the overall system interference occurs which, under the best operating conditions, may be reduced to about 2%. One difficulty with such variations is that they may not be consistent throughout the mass spectrum of the affected gas, and therefore they must be minimized as far as possible. The best method of so doing is to operate the instrument with the Nier repeller plate at +10 V. This results in the loss of some sensitivity but, on the other hand, there is now the further advantage that the ion current of any component is linearly dependent upon its partial pressure over the entire, usable range of gas pressures.

A further source of interference arises from "memory effects". Many highly polar compounds are adsorbed upon the surfaces of the instrument and are difficult to remove entirely. These slowly desorb and provide a continuous background spectrum, sometimes in considerable amount. Such compounds include: water, alcohols, ammonia, amines generally, carboxylic acids and alkynes. Several methods may be employed to minimize this nuisance; one is the use of metal rather than glass surfaces. Some alcohols are so strongly adsorbed on glass surfaces that a reliable analysis is impossible (Barnard, 1953, p. 210). Another method is to use a heated system, as materials adsorb less strongly onto hot surfaces than cold. Also the path of the material through the leak and into the source should be made as short as possible. As tap-grease is known to absorb the noble gases, its use should be restricted, and mercury cut-off valves used wherever possible. Absorbed materials such as the alkynes may be removed by introducing into the instrument a material even more strongly adsorbed to displace the original contaminant which is then pumped away. The choice of substance depends largely on the particular analytical problem. For the analysis of hydrocarbons, water and ammonia are good scavengers; for amines methanol may be used, and for alcohols an amine. In these laboratories, where the emphasis tends to be upon compounds of higher molecular weight, pyrrolidine has been found very suitable for the removal of adsorbed materials.

It follows that, for the most accurate analysis, a comparison of known and unknown mixtures of about the same composition should be made, as a means of minimizing the errors caused by interference. With this restriction, the reverse process, namely that of determining composition from peak heights, may be undertaken.

If the peak height for a given mass is denoted by H, then a series of linear equations relating partial pressures, sensitivity coefficients and the peak heights may be set down

$$\begin{array}{lcl} S_{11}p_1+S_{12}p_2+S_{13}p_3+\ldots & = & H_1 \\ S_{21}p_1+S_{22}p_2+S_{23}p_3+\ldots & = & H_2 \\ \ldots \quad \ldots \quad \ldots \quad \ldots & & \ldots \\ S_{m1}p_1+S_{m2}p_2 & = & H_m \end{array}$$

where p_m refers to the partial pressure of component m in the sample tube, H_m refers to the measured abundance of mass m in the mixture mass spectrum and S_{mn} relates to the measured abundance of the ion of mass m at unit pressure of the component n in the mixture.

This provides a set of m equations in n unknowns where, in general, $m > n$. Assuming that $m = n$, then the above series of equations can be solved by standard methods. If $m > n$ the whole array can be solved to give a best solution by the method of least squares. Both types of analysis have been attempted, and both have had their supporters. The former condition implies rejecting some rows of the matrix to reduce it to square form, and the judgement of the operator is necessary. This may have certain disadvantages but, on the other hand, it is possible to reject or give a low statistical weight to those ions which are known to be markedly temperature-dependent. For further discussion of this problem, reference should be made to other sources (Barnard, 1953, p. 202). There is, however, one extension of the method which will have a relevance to further problems discussed here. The method of measuring peak abundance ratios avoids the problem of determining total pressures (Johnsen, 1947). This method is particularly valuable for two-component mixtures, but becomes unwieldy if more components are present. Nevertheless, the method is of wide use and of great importance in the study of organic compounds. Its application depends upon the comparison of the cracking pattern of a known binary mixture with an unknown mixture of the same. Let the ratio of partial pressures in the known mixture be p_1/p_2 and in the unknown mixture be p_1^1/p_2^1 and it is the last ratio that must be determined. For the known mixture one has

$$S_{11}p_1+S_{12}p_2 = H_1$$

$$S_{21}p_1+S_{22}p_2 = H_2$$

where the symbols have their usual significance. Re-arranging

$$S_{22}/S_{21} = [H_1/H_2-S_{11}/S_{22}]p_1/[S_{12}/S_{22}-H_1/H_2]p_2$$

For the unknown

$$p_1^1/p_2^1 = [S_{12}/S_{22}-H_1^1/H_2^1]S_{22}[H_1^1/H_2^1-S_{11}/S_{21}]S_{21}$$

or on substituting

$$p_1^1/p_2^1 = \frac{[S_{12}/S_{22}-H_1^1/H_2^1]\,[H_1/H_2-S_{11}/S_{21}]p_1}{[H_1^1/H_2^1-S_{11}/S_{21}]\,[S_{12}/S_{22}-H_1/H_2]p_2}$$

The need for measuring absolute pressure has been eliminated. The accuracy of the method depends upon that of the synthetic blend.

In the early days in which the above techniques were used, the mathematical manipulation of these large matrices was cumbersome. Even when this had been successfully accomplished, the method was not always very satisfactory, since components present in only small amounts tended to be overestimated. Moreover, the analysis could not distinguish between *cis* and *trans* isomers, as in butene, because the cracking patterns of the individual isomers were so much alike. The analysis therefore yielded only the total amount of the butenes or other olefins.

When, despite limitations, it was more convenient than any other available method, a great deal of attention was given to improving it. The introduction of gas–liquid chromatography, which was relatively inexpensive and simple to operate and which could provide a quantitative analysis of simple mixtures, has largely superseded the mass spectrometer.

While this is certainly true of multicomponent mixture analysis in the oil industry, there remain other problems, some accidental, in which a familiarity with the principles of mixture analysis is necessary. One obvious application is in the analysis of a compound which is contaminated with a second component. Less frequently, but equally important, is the problem that arises when a compound, however rigorously purified, readily reverts to an isomeric mixture. Thus α-humulene, although carefully purified, readily returns to a mixture that contains some 20% of the β-isomer (having an exocyclic double bond), while the purest sample of the β-form yet reported contains about 20% of the α-isomer.

B. Fractionation

A further method of mixture analysis, which is particularly useful when there are only small quantities (1 mg or less) of material available, is that of fractionation in the mass spectrometer. It may reasonably be assumed that one component will have a higher vapour pressure than the other. In the most favourable cases, the impurity will be the more volatile. Such problems are not uncommon, as many samples contain traces of the solvent of recrystallization, and compounds having acetone, alcohol, or benzene of crystallization are occasionally encountered. One may separate the two components by fractional distillation in the heated inlet system of the mass spectrometer. The temperature of the system can be adjusted so that the impurity is either completely vaporized, or at least at such a temperature that its vapour pressure is high in relation to that of the main component. Mass spectra would then be recorded at suitable time intervals. The ions arising from the impurity would diminish more rapidly in abundance than the others and the impurity would be recognized by this. Eventually, a spectrum would be obtained in which the relative abundance of one ion to another would not change with time; this would be the required spectrum.

Such a favourable circumstance is, however, not always encountered. Often the impurity has a lower vapour pressure than the main component, or both compounds have comparable pressures. In such circumstances, only a partial separation may be effected. This is often sufficient for an analysis to be carried out, particularly if the relative abundances in the initial mixture can be determined by other methods. Several variations of the technique can be used. The material may be introduced into the mass spectrometer by total vaporization. A spectrum is taken and the least volatile part then condensed out. A second spectrum will show, by comparison with the first, which ions belong to the less volatile component, as these will differ most in the two spectra.

A further, more hazardous, method is to introduce the whole of the sample into the source and obtain a mass spectrum. Part of the sample is then pumped away; the component having the higher vapour pressure will be preferentially removed. A second mass spectrum will enable one to determine which ions belong to the more volatile substance and, in consequence, one can assign the ions to each component.

The methods so described should allow a qualitative separation of the components of the mixture and may reveal the approximate quantities of each. Analyses of this type are rather too empirical for precise computations.

The partial separation of the binary mixture having been achieved, the analysis proceeds as follows. Let the one component in the partly fractionated material have a partial pressure p_{11} and that of the second p_{12}. For the second mixture the corresponding values are p_{21} and p_{22}. If the heights of the ions are H_{m1} and H_{m2}, respectively, for each component referred to the base peak in their respective spectra and S represents the height per unit pressure of the base peak, then the peak heights at mass m in the spectra of each fraction are

$$P_{m1} = H_{m1}S_{m1}p_{11} + H_{m2}S_{m2}p_{21}$$

and

$$P_{m2} = H_{m1}S_{m1}p_{21} + H_{m2}S_{m2}p_{22}$$

where P_{m1} represents the total pressure for component m in the first fraction. Therefore

$$p_{21}P_{m2} - p_{11}P_{m2}$$

gives the spectrum of the second component. The problem, which can be settled only by trial and error, is to decide what fraction of one spectrum should be subtracted from the other. The best empirical guide is to use a factor such that many ions cancel each other out but there are no large negative peaks. This method has been successfully applied to the analysis of a mixture of *p*-methoxyazobenzene (1) and chalcone (2) (Meyerson, 1959; McCormick, unpublished observations). The method used was as follows. Two spectra corresponding to different mixtures of the two components were obtained; the ratio of the peak heights of the corresponding ions was determined, and the maximum and minimum ratios calculated.

Table XXV contains a representative sample of the ions recorded, together with the ion abundances (in arbitrary units) which are appro-

TABLE XXV

Ions from *p*-methoxyazobenzene and chalcone mixture

m/e	Mixture I	Mixture II	II/I	2·2(I) − II %	II − 0·57(I) %	(1) %	(2) %
50	249·6	358·9	1·44	15·4	17·8	16·0	18·8
51	755·5	1140·7	1·51	42·3	58·2	—	—
63	235·2	223·1	0·95	23·9	7·3	24·1	7·8
64	273·6	179·5	0·66	34·3	1·9	34·1	1·7
66	12·2	9·1	0·75	1·4	—	1·4	—
75	72·5	122·5	1·69	3·0	6·7	3·0	6·9
76	129·6	215·3	1·66	5·7	11·6	6·0	12·0
77	1504·0	2077·0	1·38	100·0	100·0	100·0	100·0
78	148·8	189·2	1·27	11·2	8·6	10·9	8·5
79	41·5	30·5	0·73	4·9	0·6	4·4	0·6
89	40·5	82·5	2·04	0·1	4·9	1·1	4·5
90	15·3	28·4	1·86	—	1·6	—	1·4
91	15·8	14·2	0·90	1·7	0·4	1·8	0·5
92	296·6	179·5	0·61	—	—	—	—
102	101·0	212·4	2·10	0·8	12·7	—	13·8
103	249·6	550·0	2·20	—	33·4	—	33·2
105	191·0	404·5	2·12	1·3	24·2	2·2	23·2
107	521·3	315·3	0·60	67·5	1·5	58·4	1·6
108	49·8	30·7	0·62	6·4	—	5·3	—
115	80·4	58·5	0·73	9·6	1·0	9·4	1·3
130	50·0	110·5	2·21	—	6·7	—	6·4
131	178·6	392·9	2·20	—	23·9	—	23·3
135	188·2	114·3	0·61	24·3	0·6	19·3	0·6
141	36·1	21·9	0·61	4·7	—	4·1	—
152	22·5	39·5	1·76	0·8	2·2	0·7	2·4
165	59·5	131·0	2·20	—	8·0	—	8·1
178	54·5	116·5	2·14	—	7·0	—	7·4
179	108·5	232·8	2·15	—	14·0	—	14·7
180	35·5	75·0	2·11	—	44·9	—	48·3
207	326·4	695·5	2·13	1·8	41·8	—	43·2
208	255·4	545·1	2·13	1·4	32·8	—	34·2
209	40·5	85·5	2·11	—	5·1	—	5·2
210	4·8	9·0	1·88	—	0·5	—	0·6
212	206·0	115·5	0·56	27·4	0·1	—	0·6
213	33·5	19·0	0·57	4·4	—	3·5	—

priate to the two mixtures for any given mass. The maximum ratio observed was 2·2 and the minimum 0·56.

Let the abundance of the ion of mass m in each mixture be represented by H_{I} and H_{II}. Further let the partial pressures of the two components be p_1 and p_2, and p_1^1 and p_2^1 in the two mixtures, respectively. Then

$$H_{I} = S_{m,1}p_1 + S_{m,2}p_2$$

and

$$H_{II} = S_{m,1}p_1^1 + S_{m,2}p_2^1$$

where $S_{m,1}$ and $S_{m,2}$ have the meanings already ascribed to them.

$$\frac{H_{II}}{H_{I}} = \frac{S_{m,1}p_1^1 + S_{m,2}p_2^1}{S_{m,1}p_1 + S_{m,2}p_2}$$

or, dividing top and bottom by $S_{m,1}$ and putting $k = S_{m,2}/S_{m,1}$,

$$\frac{H_{II}}{H_{I}} = \frac{p_1^1 + kp_2^1}{p_1 + kp_2}$$

where all the partial pressures are constant at their appropriate values during the experiment. The observed variations in the ratio of H_{II}/H_{I} must therefore depend upon variations in k. Large values of k, implying a small value for $S_{m,1}$ relative to $S_{m,2}$, will lead to the smaller ratios of H_{II}/H_{I}. Large values of this ratio will result when $S_{m,2}/S_{m,1}$ equals zero. Therefore, one may assume that the maximum ratio observed occurs when the appropriate value $S_{m,1}$ is large and $S_{m,2}$ zero. Making this assumption, one can calculate the contribution of component 1 in mixture I as shown in Table XXV above. Similar calculations also yield the mass spectrum of compound 2. In order to facilitate comparison with the spectra of the pure components listed in the last two columns, the figures in the last four columns are all referred to their most abundant ion as the base peak. It will be seen that, with one or two exceptions, the agreement between the derived and authentic mass spectra are good. The agreement is, in fact, comparable to that observed when running the same material in similar instruments of the same design. The partial pressure of each component can now be determined and this will allow a quantitative analysis of the initial mixture. Two methods can be used. All the information available can be employed and a least squares method used to obtain the best result; or judgement may be used in selecting the equations to be used. This last method is employed here.

Reference to Table XXV shows that, for the ion $m/e = 141$, compound 1 makes no contribution to the peak height of either mixture. The same is true of the ion $m/e = 180$ for compound 2. Reverting to the original equations.

$$S_{141,1}\,p_1 + S_{141,2}\,p_2 = H_{\mathrm{I}}$$

and

$$S_{180,1}\,p_1 + S_{180,2}\,p_2 = H_{\mathrm{II}};$$

where H now refers to the peak height of a mixture and S represents the peak height per unit of pressure of the pure component. Now from the derived spectrum $S_{141,2} = S_{180,1} = 0$. Substituting the appropriate values and re-arranging, the partial pressures for mixture I are $p_1 = 0{\cdot}74$ and $p_2 = 0{\cdot}26$ while, for mixture II, $p_2 = 0{\cdot}45$ and $p_2^1 = 0{\cdot}55$.

The analysis is more troublesome for a three-component system. Attempts are being made to solve this problem and to extend the analysis to even more complicated mixtures. So far, the results obtained are not sufficiently refined to recognize the cracking patterns, and this part of the problem is being further investigated.

A similar investigation has been reported in the degradation of pithecolobine (Orr and Wiesner, 1959), in which a mixture of acids was discovered by just such an experiment. The degraded material was introduced into the mass spectrometer, when a series of ions $m/e = 200$, 214, and 228 were observed. On raising the temperature of the material and re-examining the spectrum, it was found that the relative abundance of these ions to one another had changed. The ion $m/e = 200$ became less abundant with respect to both the others. A further rise in temperature and a re-examination of the spectrum showed that, not only had the abundance of the ion $m/e = 200$ diminished with respect to the others, but that $m/e = 214$ was also diminishing with respect to $m/e = 228$. This reveals the presence of three distinct carboxylic acids, which were recognized as lauric, tridecanoic and myristic acids.

A further ion $m/e = 185$ was also present in moderate abundance. It was found to remain in very nearly the same ratio to the ion $m/e = 228$ throughout the whole temperature range investigated. It is, therefore, a fragment ion which is mainly, if not entirely, derived

$$C_{13}H_{27}\overset{+}{C}O_2H \rightarrow C_{11}\overset{+}{H}_{21}O_2 + \text{residue}$$

from myristic acid. Other ions of low abundance which are present at the higher-mass end of the spectrum may be analysed similarly.

The second method is to examine the spectrum of the mixture at a series of temperatures which, if possible, should cover a sufficient range to have all of the components in the vapour phase at least at one temperature (see Section D, p. 141).

A further analytical method depends upon a simplification of the mass spectrum by running the trace at an electron beam energy which is greater than the ionization potentials of some molecules but below that of others. Such a method has previously been discussed (p. 13). In very favourable examples, the trace may contain all the parent molecular ions but, because the energy of the electron beam was less than that needed to produce fragment ions, only these. This provides a method of enumerating the parent ions although, because the characteristic cracking patterns are absent, no structural information can be obtained.

In other circumstances, by a suitable choice of energy, it may be possible to ionize the unsaturated molecules, which have a lower ionization potential than the saturated. Then, the molecular ions present may be grouped in two classes and a qualitative assessment made of each.

The practical method of obtaining the results is to record a spectrum at progressively lower electron energies until this value is less than the threshold for any component present. From these charts one can not only determine approximately the ionization potential of many molecular species, but also the rate of change of the abundance of any ion with electron beam energy. In the neighbourhood of the ionization potential, the rate of change of abundance of the molecular ion with the energy in excess of this potential is roughly the same for all singly charged positive ions. Fragment ions, particularly those formed by multiple cleavage, may not only require more energy before they appear, but may also have a quite different dependence upon the excess energy. Care should be taken in the use of this analytical method, however, as very often facile though complicated processes may occur.

The method has been successfully employed in the analysis of an

TABLE XXVI

Spectrum of oils from *Leptospermum scoparium*

m/e (Obs.)	Probable formula	*m/e* (Calc.)	p.p.m.
204·18734	$C_{15}H_{24}$	204·187792	4·5
202·17105	$C_{15}H_{22}$	202·172142	−10·9
200·15503	$C_{15}H_{20}$	200·156492	−14·6
159·11686	$C_{12}H_{15}$	159·117370	5·1
134·11172	$C_{10}H_{14}$	134109595	−21·2

extract of the essential oils of *Leptospermum scoparium* (H. C. Hill, personal communication). Mass spectra of the mixture were obtained at low electron energies in order to detect parent molecular ions and exact mass measurements were made upon many of them. A few of the results are shown in Table XXVI, in which the precise masses and molecular compositions are recorded.

The exact mass measurements reported in Table XXVI were made at an electron energy of 8 eV. The systematic error of the instrument for precise mass measurement was 4·83 parts per million. The measurements are not of a particularly high precision, but even so there is no doubt as to the molecular constitution. Chemical tests had already shown that no nitrogen-containing material was present. One component is, therefore, a fragment ion which underlines the warning already given.

C. Additional Methods

In addition to all the examples so far discussed, there is a group of problems which require other special techniques or applications. These will now be discussed.

I. EFFUSION

One such method was first used in an investigation of the spectrum of HD (Friedel and Sharkey Jr., 1949). In order to measure the relative abundance of D^+ as accurately as possible, an estimate of the amount of H_2^+ was needed. This was obtained by measuring the rate of diffusion of molecular hydrogen under standard conditions, as well as that of HD ($m/c = 3$); a graph was constructed (Fig. 17).

The measured rate of decrease in the height of the ion ($m/e = 2$) could then be placed on this line, which would give the relative amounts of H_2 and HD in the total abundance of the ion. (Such a procedure would now be unnecessary, since instruments which have a resolving power of about 8 parts in 10^4 would separate the two ions.) The method may also be applied to multicomponent mixtures.

In some exceptional cases, the organic molecule does not give rise to a parent molecular ion, nor in some examples can a $(P+1)^+$ ion be produced by a secondary collision process. In such circumstances, an effusion method may be used. The unknown substance is allowed to diffuse out through some porous material, the rate of effusion into the source being determined by the increase in abundance of any given ion formed by electron impact. This rate may then be compared with that of a compound of known molecular weight. Provided that the initial pressure of the known and unknown materials are the same, the simplest case, the molecular weights are inversely proportional to the square of the times of effusion. Certain other precautions must be observed, and it is important that the vapours used in the experiment neither dissociate, associate, nor become strongly adsorbed on the walls of the instrument during the course of the experiment.

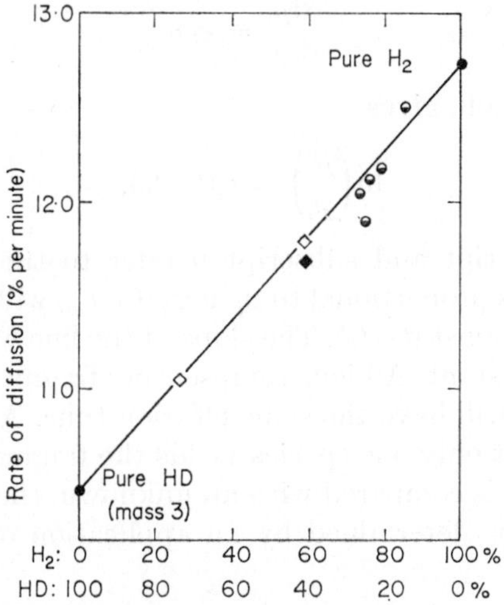

FIG. 17. Rate of diffusion plot to determine H_2 impurity and relative abundance of D^+. ○ = HD sample 1. ◒ = Synthetic blend H_2 and HD sample 1. ◇ = HD sample 2 (highest purity). ◆ = synthetic blend H_2 and HD sample 2.

The more general application is as follows. If $\mathscr{H}_k$ is the peak height of an ion measured in arbitrary units then

$$\mathscr{H}_k = H_{1,k} + H_{2,k} + H_{3,k} + \ldots$$

where $H_{1,k}$, $H_{2,k}$, etc., represent the individual contributions to the height of the peak $\mathscr{H}_k$ from the different components in the mixture.

Now if $H_{i,k} = S_{i,k} p_i$ where $S_{i,k}$ is the sensitivity, i.e. the function that relates the partial pressure of ith species to the height of the ion beam $H_{i,k}$, and p_i is the partial pressure of the ith species, then

$$\mathscr{H}_k = S_{1,k} p_1 + S_{2,k} p_2 + S_{3,k} p_3 + \ldots$$

The material is pumped out of a reservoir and through an orifice. If the pressure on the sample side of the orifice be p_i and on the low pressure (ion chamber side) p_w, then the rate of effusion, which is proportional to the pressure difference, is

$$-\frac{dp}{dt} = c_i(p_i - p_w);$$

where c_i the constant of proportionality, is the effusion constant relating to the ith species.

If the low pressure is considered as zero (a condition which is a reasonable approximation since ion sources usually operate at pressures of 3×10^{-6} torr or less), the equation becomes

$$\frac{-\mathrm{d}p_i}{\mathrm{d}t} = c_i p_i$$

which, on integration, gives

$$\ln\left(\frac{p_i^0}{p_i}\right) = c_i(t-t_0),$$

where the superscript and subscript 0 refer to the initial condition. Now, because H_i is proportional to p_i, $\log(H_i/H_{ik})$ will yield a rectilinear graph if plotted against $(t-t_0)$. The slope of the line will be proportional to the effusion constant. All ions representing fragments from the same molecular species will have the same effusion time. A linear plot, moreover, confirms that only one species yields the fragment. If the rate for a known fragment is compared with an unknown, then the unidentified component may be determined by an application of Grahams law of diffusion.

2. LOW-ENERGY SPECTRA

The use of low-energy electron beams has also been employed in several special problems. The method is not without its dangers, however. These low energies are often employed in the determination of the isotopic enrichment in any molecule. The merit of such spectra is that, since the electron beam energy is close to that of the parent ion, the spectrum is essentially one of the parent ion only, uncomplicated by any fragmentation pattern.

There are several possible sources of error to be guarded against. It is important that the electron beam energy should remain constant during the analysis, as the effective collision cross-section of the substance varies markedly with electron energy in the neighbourhood of the ionization potential. Moreover, it is obviously very important that ion–molecule collisions, which would produce an ion at $(\mathrm{P}+1)^+$, do not occur.

The first of these may best be checked by carrying out several measurements. The second is more troublesome. The effect is not reproducible, as the incursion of the $(\mathrm{P}+1)^+$ ion arises from a secondary process, being dependent upon the basicity of the compound, upon the vapour pressure of the material and upon the potentials existing in the ion source of the mass spectrometer. The safest way of detecting this kind of ion is to measure its abundance at different partial pressures of the substance. Provided that the pressures are known, the contribution to the ion fragment, which depends upon the square of the pressure and is the product of ion–molecule collision, may be removed to yield the amount showing a first order pressure dependence, the true isotopic ion

D. Latent Heat of Change of State

A final determination which is sometimes useful is the measurement of the latent heat of change of state. As already mentioned (Chapter 2, p. 27), the Clausius-Clapeyron equation relates the saturated vapour pressure of any compound to the latent heat of vaporization if the substance is a liquid, or of sublimation if it is a solid. Any function which is proportional to the saturated vapour pressure is equally serviceable, and the constant of proportionality need not be known.

Consider a compound in the ionization chamber of the mass spectrometer. Provided that the pressure of the vapour in the chamber is proportional to the saturated vapour pressure, which will be true if the material is not absorbed or condensed and the pumping speed is adequate, the abundance of the ion is known to be proportional to its vapour pressure under fixed operating conditions. Therefore, the abundance of the parent molecular ion may be plotted against the reciprocal of the absolute temperature as a means of determining the latent heat of change of state.

The method may also be employed using a fragment ion as a measure of the compound's vapour pressure. It is thus possible to determine the latent heat of change in substances which do not give parent ions. Also, one can determine that quantity in impure compounds provided that the ion being measured derives solely from the material under investigation, with no contribution from any impurity. This method has been exploited in a mass spectrometric determination of the vapour pressure of propane (Tickner and Lossing, 1950, 1951).

In favourable circumstances it may be employed for the recognition of mixtures, although the experiment is a long and rather tedious one. Some circumstances might require such an analysis. A mixture of isomers as may commonly be found amongst the monocyclic monoterpenes can have all the major fragment ions in common. The presence of a mixture will be revealed if a plot of the peak height against the reciprocal of the absolute temperature is made. The graph will not be rectilinear, particularly if the latent heats of vaporization are markedly different. Such a result is shown in Fig. 18. This shows the plot of a synthetic mixture of camphor and dimethylaniline as well as those of the pure components. It will be apparent that the plots for the mixtures are curved, although only slightly. The successful operation of this technique

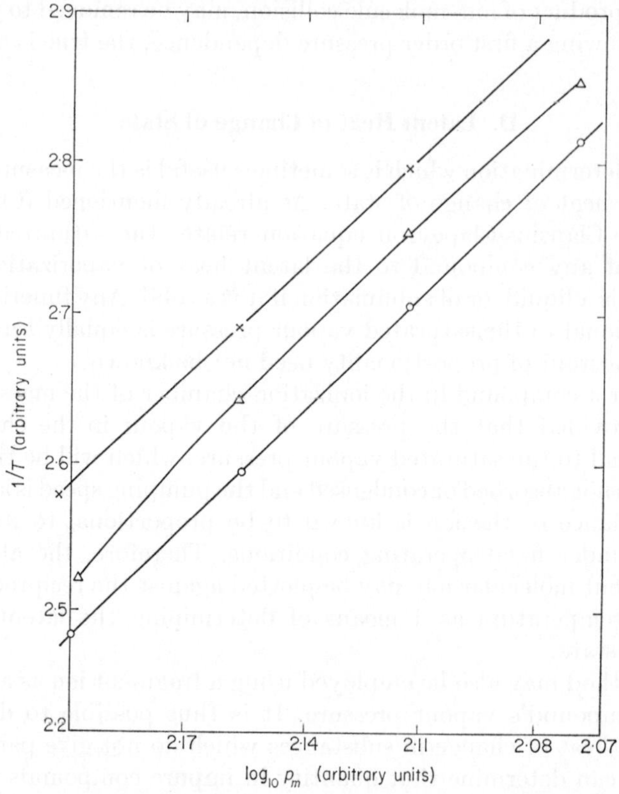

FIG. 18. Plot of peak height against reciprocal of the absolute temperature for a typical synthetic mixture. ×, Camphor. ○, Dimethylaniline. △, Mixture.

demands good temperature control and great precision in measurement. By such an analysis, it may be possible to pick out ions which belong either wholly to one species, or at least which make a major contribution to one component only.

CHAPTER 7

The Analysis of Spectra

A. The Analysis of a Spectrum

In view of what has been reported already it is clear that hard-and-fast rules cannot be given, at least at this time, for the analysis of spectra. The following procedure is set out for guidance in the matter. There may exist, and indeed must do, many short cuts which will depend upon the skill and experience of the analyst and the assistance that may be obtained from other sources. In the present scheme, there has been no recourse to analytical aids other than physical properties which in favourable circumstances will assist the analyst in positioning substituent groups.

1. PROCEDURE

1. Obtain the spectrum and determine the molecular weight. Measure the abundance of the parent molecular ion as a percentage of the base peak. In the uncommon event of a parent molecular ion not being present obtain a $(P+1)^+$ ion at higher gas pressure or determine the molecular weight by an effusion technique.
2. From the spectrum, or by exact mass measurement, determine the molecular composition of the substance, including the number and nature of the heteroatoms present.
3. The compound should be reduced, purified and the spectrum re-run (see Appendix 1, p. 190). The molecular weight may be redetermined if necessary. The increase in molecular weight will disclose the number of double bonds which were present.
4. Exact mass measurement, or isotope analysis, will reveal the presence of hetero-atoms which still exist after reduction.
5. Pyrolysis in the sample admission system may next be tried, but only if hetero-atoms are present.

B. Hydrocarbons

1. ALIPHATIC

The molecular weight obtained shows a saturated alkane. Determine the structure either from comparison with known spectra, or by the methods already outlined (Chapter 3, p. 43).

The molecular formula indicates possible unsaturation. The structure may be recognized from past experience or from its similarity with a known spectrum. Alternatively, reduction will indicate the number of double bonds and the molecular structure of the resulting alkane may be determined.

2. AROMATIC

Resistance to reduction is shown by aromatic nuclei and unless the hydrogenation is vigorous these will be unaffected. Such nuclei are readily detected from the base peak which is often 91^+ or of the form $(91+14n)^+$; also the further elimination of an acetylene unit (26 mass numbers) may occur from the base peak.

C. Epeisactic Hetero-atoms

The nature and number of these atoms will often give a good clue as to their identity.

1. SINGLE OXYGEN

If the molecular ion contains one oxygen and no double bond equivalents, it must, in this classification, represent a hydroxyl or ether group. If the molecular ion contains one double bond which is reduced to yield an alcohol, the original function was a carbonyl. Location of the hydroxyl group will complete the solution. If the derived alcohol is primary, the original molecule was an aldehyde—otherwise a ketone. The ketonic function may be located by known methods (Chapter 4, p. 86).

In a previous chapter, when an approximate classification of the hetero-atoms was made, it was found convenient to regard the methoxyl as a single entity. Accordingly, it is included in the present class.

This function is probably the most difficult to recognize amongst all those that are commonly encountered. Often, fragmentation in the aliphatic portion may be extensive, whilst the cleavage of the methoxyl is small and insignificant. Moreover, when such fragmentations do occur, it is not always possible to distinguish between this group and a hydroxymethyl. For these reasons, the easiest method of analysis is probably to recognize the nature of this group by physical methods. It is well known that alcohols are rather high boiling for their molecular weight; methyl ethers of a similar mass have a markedly lower boiling point. Since the molecular weight is obtained from the cracking pattern, some decision might be made upon this basis alone. Further chemical confirmation may be obtained from the reactivity of the alcohol. This compound may be methylated by standard methods to convert it to the methyl ether or to the silyl ether, which has been reported to be even more volatile than the corresponding hydrocarbon (Sharkey Jr., Friedel and Langer, 1957), reactions which cannot occur with the methyl ether.

2. TWO OXYGENS

Several possibilities exist, for example the following. (i) Saturated aliphatic or alicyclic structures. Two alcohol groups which may be positioned as in Section B, 1; (ii) One double bond equivalent (one carbonyl and one hydroxyl) proceed as in Section B, 1. Carboxylic acid or ester. Not affected by mild reduction. Proceed as under Sections H, J and K of Chapter 4. A peroxide is also possible although less commonly encountered. Reduction will remove one oxygen. Proceed under the appropriate section.

Acetals and ketals are found in this group. These are usually easy to recognize by their small parent molecular ions and the fact that on acid-calatysed reduction, they yield an alcohol.

3. THREE OXYGENS

Here many possibilities arise. Only three of the commonest will be discussed.

(i) One double bond equivalent not reduced under mild conditions indicates a hydroxy acid or ester. The presence of an acid may be detected by the usual chemical tests. Positioning of the two groups may be done by the methods discussed above.

(ii) One double bond equivalent which can be reduced under mild conditions indicates a carbonyl function. Reduction will identify this. Recognition of the hydroxyl as a primary or secondary alcohol will show the nature of the original function. Proceed as above.

(iii) Trihydroxy compound. No change in oxygen functions upon reduction. Boiling or melting point is usually high in relation to the molecular weight.

4. SINGLE NITROGEN

If the molecule is saturated, the nitrogen can only be present as an amine. Recognition of this should be fairly easily made (see Chapter 4, p. 77). If one double bond equivalent is present, the compound is an imide. The location of the function is most conveniently done on the reduced compound. If again there exist two double bond equivalents directly attached to the nitrogen, the compound is a nitrile. These can often be identified by their characteristic cracking pattern without the need for reduction (Chapter 4, p. 83).

5. TWO NITROGENS

(a) No double bond equivalent

A diamine. Analysis may be attempted as before or aided by chemical manipulation. Treatment with nitrous acid will yield an alcohol for any primary amine. The mass spectrum of the product will contain one oxygen for each primary amine and a nitrosogroup for each secondary. There is no change for a tertiary amine. The hydroxyls may be fairly readily located which assists in determining the original structure.

(b) One double bond equivalent

This group may represent an amine and imide or much less commonly an aliphatic diazo-alkane. Reduction of the first of these leads to a diamine, and one can proceed as above.

(c) Two double bond equivalents

This may represent one amine and one nitrile group. The positioning of both functions may, in this instance, cause some difficulty, unless the nature of the amine is first determined. Perhaps the best systematic method is to reduce the nitrile to an amine and then treat it with nitrous acid.

A mass spectrum and determination of the molecular constitution will reveal the groups present originally. If the new compound contains: (a) no nitrogen but two oxygens, the original contained a nitrile and primary amine; (b) one nitrogen and two oxygens, the original had a nitrile and secondary amine; (c) one nitrogen and one oxygen, the initial compound had a tertiary amine as well as a nitrile.

The correctness of these deductions can be tested by confirming that a primary alcohol is present in the reduced compound after treatment with nitrous acid, a consequence of the reaction sequence

$$R{-}C{\equiv}N \xrightarrow{4H} RCH_2NH_2 \xrightarrow{HNO_2} RCH_2OH + N_2 + H_2O$$

$$RCH_2O\overset{+}{H} \longrightarrow \overset{+}{R} + CH_3O\cdot$$

$$(P\overset{+}{-}31)$$

A second possibility will be two imides. Reduction will yield two amines. Treatment with nitrous acid and determination of the molecular constitution of the new compound will indicate whether the imides were substituted or not.

6. THREE NITROGENS

Many possibilities exist of which two only will be discussed.

(i) Three nitrogen atoms having no double bonds. Three amines must be present. Treatment with nitrous acid and a new determination of the molecular formula will aid the classification of these groups. The usual methods will then enable one to locate the nitrogens in the original compound.

(ii) Three nitrogen atoms being associated with four double bonds. Several possibilities here exist, which are not resolved by treatment with nitrous acid and an examination of the products, although this method does reduce the number of possibilities. For instance a system possessing either two nitrile groups and a primary amine or one having one nitrile and two imides would yield a trihydroxy compound upon reduction followed by treatment with nitrous acid. Such a complex system is troublesome to analyse and would require a detailed consideration of the original cracking pattern. The trihydroxy derivative would be most useful in positioning the substituents.

7. SULPHUR

The chemistry of sulphur containing compounds is similar to that of oxygen.

(a) One sulphur—no double bond

The compound is a thiol, within the limitations imposed in the groups under discussion. Positioning along the molecular structure would be carried out by the methods reported earlier in this chapter (p. 144).

(b) One sulphur—one double bond

Reduction would establish whether the C=S was terminal or not. Care is needed in these compounds that the sulphur is not eliminated.

The methods used for locating a ketone function will operate less clearly, in a thioketone.

(c) Two sulphur atoms—no double bond

This represents a dithiol and the groups may be positioned by the standard methods already used. A persulphide would be detected by mild reduction which would remove one sulphur atom.

(d) Two sulphur atoms—one double bond

This may be either a thiol and C=S attached to separate carbons, the sulphur analogue of a carboxylic acid, or its ester. The first possibility may easily be recognized by mild reduction. The second and third have not been studied as a class and their characteristic fragmentation patterns are not known.

(e) Three sulphur atoms

So many possibilities exist that it is not practicable to cover them all. Mild reduction will help solve the problem in that it will assist in classifying and locating the substituents.

8. HALOGENS

(a) One halogen

This is most readily detected by removing the halogen and comparing the mass spectrum of the alkane with that of the original compound. This procedure is usually unnecessary with fluoro compounds since it is well known that fluorine closely resembles hydrogen in properties upon impact. It is therefore only necessary to recognize the alkyl ion fragments containing the fluorine atom for the structure of the molecule to be determined.

(b) Two halogens

This problem may be treated in the same manner as that immediately preceding. Except for fluorine, the problem may be simplified by hydrogenating out the halogen and comparing the original spectrum with that of the alkane now obtained.

(c) Further halogens

The introduction of further halogens does not affect the general principle nor the recommended procedure.

D. Mixed Hetero-atoms

This class must naturally cover a very big field of organic chemicals and an exhaustive treatment is not possible. Fortunately, the analytical procedure can be systematized, and the labour reduced. Even so, only some of the commoner functional groups can be examined here.

(i) There is a large class of compounds possessing two or more hetero-atoms which represent two or more separate functional groups. Ignoring the possible effects of the interaction of these, they may be considered as two separate epeisactic hetero-atoms. The general procedures already designed may be suitably modified to cover such problems. The molecular weight and constitution can be obtained, the substance hydrogenated and re-examined. The reduced product is then treated with nitrous acid if nitrogen was originally present. The mass spectrum is re-taken. From this, the molecular weight and composition can be determined and the nature and position of the original substituents derived.

Occasionally, the problem may be complicated by specific difficulties. A molecule containing one nitrogen and two oxygen atoms may be, among other possibilities, an amino acid. High-temperature treatment of this yields a diketopiperazine—not the original substance. Difficulties of this kind can always be avoided by determining the boiling or melting point of the original material before mass spectroscopic examination. In this instance, such a measurement would reveal the intractable nature of the material which, in turn, suggests the use of a derivative, e.g. an ethyl ester.

(ii) Of more interest is the structure in which the hetero-atoms are themselves bonded together to give a more elaborate substituent group.

1. ONE NITROGEN AND ONE OXYGEN

One nitrogen and one oxygen may be joined to each other in more than one way and, accordingly, the definitions must be narrowed somewhat. Again we shall assume when double bonds are considered that they are between the hetero-atoms or the carbon to which these are attached. Even so, we now have two types of multiple bonding, either between the hetero-atoms, a nitroso group in this instance –N=O, or an oxime –C=N—OH. To systematize the procedure, we shall first consider bonding between the hetero-atoms and secondly bonding to the adjacent carbon atom.

(a) NO, NOH

The two groups here discussed can be easily distinguished on the original cracking patterns. The nitroso group is frequently eliminated as such, while the limited evidence of the mass spectrometry of oximes suggests that only the hydroxyl is lost.

Reduction of either will yield an amine with the loss of the oxygen atom although care must be taken that a re-arrangement of the oxime does not occur. After reduction, treatment with nitrous acid leads to a hydroxyl group which is recognizably a primary hydroxyl, from the cracking pattern. The hydroxyl can be used to locate the position of the original substituent.

(b) –CNO, –NCO

Little study has been made of organic cyanates or isocyanates. The reported example physostigmine shows that a cyanate group is readily lost as an entity, and this may well be true of isocyanates also. The two groupings may be distinguished because upon reduction the cyanate will lose the oxygen on becoming the primary amine whilst the isocyanate would be reduced to a secondary amine. In practice, the isocyanate is much less stable and may be hydrolysed to an amine, which has one carbon less than the original compound. The amine may be converted to the primary alcohol, or nitrosamine and the position in the original compound located.

(c) NO_2

This grouping is characteristic of aliphatic nitrites, or nitroparaffins. Reference to the discussion on cracking patterns (Chapter 4, pp. 84–5) shows that the patterns associated with these two types of substituent are quite distinct. Even if this were not so, reduction of the two groups leads to different products which may be easily distinguished by mass spectrometry. The nitro-paraffins may be reduced to the corresponding amines or oximes which can in turn be made into the primary alcohols, whilst the nitrite reduces directly to the alcohol. Again the alcohols may be used to position the original functional group.

(d) $-NO_3$

Alkyl nitrates are relatively uncommon. They may be easily recognized by the loss of the nitrate as a single entity. Positioning the group along the aliphatic skeleton is a more difficult problem.

(e) $RCONH_2$

Among the common compounds containing both nitrogen and oxygen are the acid amides. The mass spectrum of such amides yield a molecular weight and structural features. Some indication that one is examining an acid amide may be obtained from the melting point of the compound, which is relatively high for the observed molecular weight. The amide may, of course, have alkyl substituents attached to the nitrogen and these, as already discussed in Chapter 4 (p. 83), will have distinctive cracking patterns. A detailed consideration of the fragmentation pattern usually reveals substituents upon the nitrogen. However, this problem may be readily resolved by a vigorous reduction of the amide, which will yield the appropriate amine. Primary amides will yield primary amines which may in turn be converted to the primary alcohols. Secondary amides become secondary amines which are converted to *N*-nitroso derivatives on treatment with nitrous acid. The tertiary amine is not affected by nitrous acid. Once the character of the amide has been determined, it is relatively easy to interpret the original fragmentation pattern.

2. SULPHUR AND OXYGEN

(a) SO

Sulphoxides are uncommon. They may be identified by their tendency to cleave at the carbon–sulphur bonds. The ion $m/e = 48$ (SO^+) is also abundant. As a consequence of this form of cleavage, it is fairly easy to determine the length of the two alkyl substituents, but the structure of the entire molecule is more difficult to deduce.

(b) SO_2

This grouping is characteristic of sulphones and also of sulphinic acids or esters. Compounds in the former category show a behaviour similar to the sulphoxides in that both alkyl groups are easily cleaved, and the ion $m/e = 64$ (SO_2^+) is very abundant. Again, while it is a simple problem to determine the lengths of the alkyl chains, the determination of their structure is difficult.

Free sulphinic acids are unlikely to be sufficiently volatile for introduction through the conventionally heated source, and so far, there has been no extensive study of the sulphinic esters.

(c) SO_3

Sulphonic acids or esters are rather unsuitable for mass spectrometric study. The free acids are not sufficiently volatile, while experiments with esters indicate that they are either not stable enough or, occasionally too involatile. Even when these obstacles have been overcome by special methods, it is difficult to interpret the mass spectrum.

(d) CNS, NCS

The series of alkyl thiocyanates has not been extensively investigated. Consideration of their structure suggests that cleavage between the acid and the alkyl group would occur easily.

$$\overset{+}{\mathrm{RCNS}} \rightarrow \overset{+}{\mathrm{R}} + \mathrm{CNS}$$

Isothiocyanates have recently been reported (Budzikiewicz, Djerassi and Williams, 1964). Two fragmentations are reported

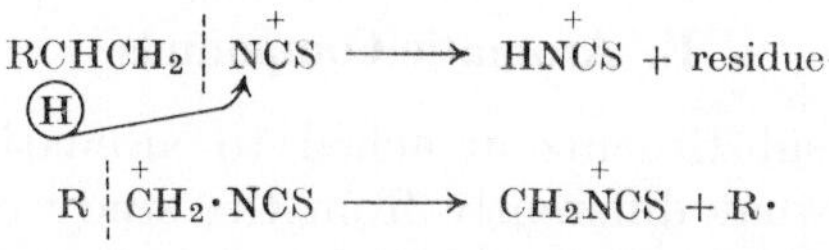

The former is the most important process for ethyl isothiocyanate, but becomes progressively less so along the homologous series. The

latter process is present in the spectra of all straight chain isothiocyanates.

Reduction with lithium aluminium hydride would distinguish between the two series, since the thiocyanate would yield a primary amine while the isothiocyanate would become a secondary amine. The primary amine may be converted to the alcohol and the other to the *N*-nitrosamine.

3. HALOGENS

(a) COX

The commonest combination of oxygen and the halogens is as the appropriate acyl halide. These may be easily recognized in the original cracking pattern. The complete group may be detached or the halogen atom alone removed. Positioning along the alkane skeleton is not very easy. One method would be reduction to the corresponding aldehyde by a Rosenmund reduction and the carbonyl further reduced to the alcohol. Alternatively, the acyl halide might be reduced in one step to the alcohol, by means of lithium aluminium hydride.

(b) CSX

As a class they do not seem to have received a systematic investigation. By analogy with the acyl halides, the halogen should readily be lost. Loss of the complete CSX group should also occur. The positioning of the substituent along the carbon chain will again be difficult.

4. ANHYDRIDES

Comparatively little systematic study has been made of these. They can easily be recognized from their cracking pattern by the presence of an abundant acyl ion. The composition of which is more readily determined with double-focusing instruments.

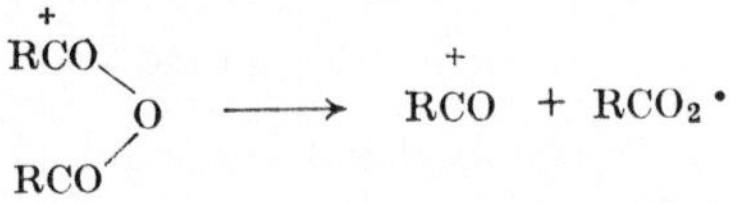

E. Aromatic Compounds

The study of substituents attached to aromatic nuclei must be approached somewhat differently from the study of purely aliphatic compounds. Also, the usual classification of aromatic and aliphatic substituents must be modified. There is much experimental evidence, (see Chapter 3, p. 67), to suggest that the favoured centre of fission is at the bond β to the aromatic ring. Accordingly, compounds such as benzyl alcohol, which would normally be classified as an aliphatic alcohol whose chemical properties it so much resembles, must be included here, since the fragmentation pattern of benzyl alcohol is distinctively different from alkanols. Therefore, there are two classes of compound which are grouped together for convenience, namely those in which the substituent group is directly attached to the aromatic nucleus, as in aniline, and those which contain a single carbon between, as in benzylamine.

1. ONE OXYGEN ATOM

(a) –OH

Oxygen may be directly attached, as in phenol. In these circumstances, the common forms of elimination are as neutral carbon monoxide, the neutral aldehyde group or its ion. It is desirable to confirm the presence of a free hydroxyl by some suitable test, because diaryl ethers often give the same sort of elimination.

(b) –CH_2OH

This group is often characterized by the loss of the hydroxyl only. Sometimes the entire hydroxymethyl group is lost but does not yield an abundant ion.

(c) –CHO

The aldehyde group directly attached to an aromatic centre has the unusual and easily recognizable property that the $(P-1)^+$ is even more abundant than the prominent parent molecular ion. Moreover, this structure may be reduced by mild reducing agents, and the resulting alcohol examined.

Since the monovalent functional group already contains carbon, the homologue –CH_2CHO will behave as an aliphatic substituent, even under the present specialized definition.

(d) –CO

The doubly bonded carbonyl may be directly joined to aromatic rings as in benzophenone or fluorenone. It may be bonded to an aromatic group on one side only as in acetophenone, directly to an aromatic on one side and to a benzyl on the other as in benzyl phenyl ketone, or to two benzyl groups as in dibenzyl ketone etc. The behaviour of these various compounds will vary somewhat under electron impact, depending upon the environment of the carbonyl function. Purely aromatic carbonyl compounds, such as benzophenone and fluorenone, lose carbon monoxide readily. The former also shows some evidence for fission to yield benzoyl ($m/e = 105$) or phenyl ($m/e = 77$) ions. Fluorenone naturally shows no such tendency. Compounds such as acetophenone lose a methyl and less readily yield acetyl ions. The probable ions from the other two compounds are also obtained, namely phenyl, benzyl, benzoyl, and phenacyl ions from benzyl phenyl ketone and phenyl and phenacyl ions from dibenzyl ketone. Conclusions of this kind are readily confirmed, since, on reduction, the corresponding alcohols are obtained. Their behaviour may be examined as discussed previously.

2. TWO OXYGEN ATOMS

Two oxygens may refer to two functional groups of the same or different type, commonly attached to distinct aromatic centres. In general, the properties of each substituent will be separately displayed and, apart from the problem of orientation (an exceedingly difficult if not impossible task on mass spectrometric evidence alone), the problem is no more difficult than that of a single constituent.

Certain accidental groupings of simple functions may lead to special difficulties in some circumstances, particularly if the molecule is a mixed aliphatic aromatic compound. Even 4-hydroxybenzaldehyde upon electron bombardment readily loses first one and then a second carbon monoxide, which would lead to the suspicion that there are two carbonyl functions in the molecule, or even possibly two hydroxyl groups on the aromatic nucleus, if the formula were not known. This difficulty is easily corrected by reducing the original compound, when the uptake of two hydrogen atoms and the presence of a primary alcohol in the spectrum of the reduced material proves the original molecule to have contained a carbonyl group. Since only two hydrogens were taken up, there must have been one carbonyl function and hence one hydroxyl or alkoxy group present originally. The mass spectrum of the first compound showed the loss of two carbonyl groups. One only is lost from the reduced product. Therefore this oxygen function must be phenolic, the only system which will eliminate carbon monoxide in the group being considered. The molecule therefore contains a phenoxy grouping and an aldehyde grouping.

$-CO_2-$

More commonly the presence of two oxygens indicates an aromatic acid or ester. The fragmentation of the acid occurs only to a small extent and is represented by the loss of a hydroxyl and the formation of an aryl ion. Fissions encountered in the aryl esters have already been reported (Chapter 4, p. 88).

3. THREE OXYGENS

As mentioned previously, the presence of three oxygens may be accounted for by a wide variety of multifunctional molecules. Only two classes are discussed here, per-acids and anhydrides.

(a) $-CO_3H$

Per-acids have not been studied to any extent, but there is little doubt that the well-established β-bond fission will occur to yield the aryl residue

$$\overset{+}{Ar}CO_3H \rightarrow \overset{+}{Ar}CO + O_2 + H$$

Re-arrangement may occur.

(b) Anhydrides

This group obviously covers several categories of anhydride. The simplest is probably the purely aromatic (in the strict sense) anhydrides formed by monobasic acids. In this case fragmentation is as follows

$$ArCO\overset{+}{O}OCAr' \rightarrow \overset{+}{Ar}CO + Ar'CO_2 \text{ or } Ar'\overset{+}{C}O + ArCO_2$$

with the possible further decomposition of the carboxylate ions which may also be formed, to give Ar^+ or $(Ar')^+$.

Mixed aryl anhydrides of the form $ArCH_2COOOCCH_2Ar'$ yield further products. The following scheme covers most of the ions likely to be formed.

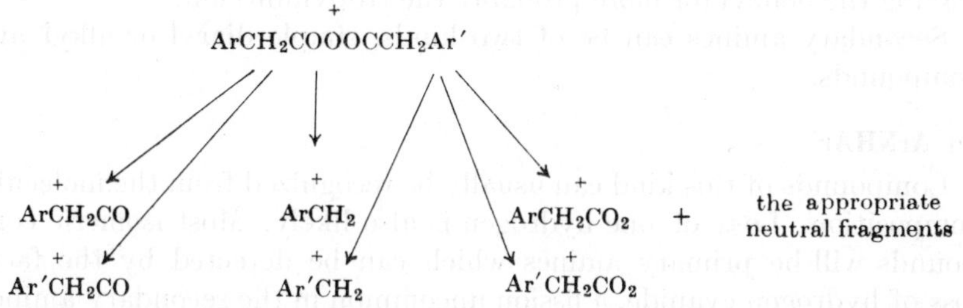

In addition, since the structure contains hydrogens which are β-bonded to the aromatic nuclei, the $(P-1)^+$ ion may be expected to be abundant.

Mixed alkyl aryl anhydrides will have the characteristics of the aliphatic moiety already discussed (see Chapter 7, p. 152). Superimposed upon this will be the fragmentation from the aryl portion which, as stated above, will be mainly the aryl ion.

4. NITROGEN CONTAINING COMPOUNDS

The nitrogen may be present as a primary, secondary or tertiary amine. Decision as to which, amongst these possibilities, is the most likely is very easily obtained by chemical methods. If necessary, the correct conclusion may be reached by a careful examination of the cracking pattern. Should the system be treated with nitrous acid and heated, the primary amine will be converted to a phenol, the secondary amine to a nitroso and often the tertiary amine to a *C*-nitroso derivative. Therefore, if the mass spectra of the original and the treated product are examined, for a primary amine the nitrogen will be replaced by an oxygen and for a secondary amine the new compound will contain a further nitrogen and oxygen as is also true of many tertiary amines in which a phenyl group is directly attached to the nitrogen. The distinction between *N*-nitroso and *C*-nitroso will be apparent in the cracking patterns. The *N*-nitroso should readily lose NO.

Positioning of the amine group may be more troublesome since, unless this group is very close to another, it is very difficult to decide the orientation. In the simpler examples, where the amine is the only substituent, the problem is somewhat easier.

(a) **$ArNH_2$**

The primary amines such as aniline tend to lose HCN rather easily, so that there is an abundant ion at $(P-27)^+$.

(b) **$ArCH_2NH_2$**

Compounds such as benzylamine tend to lose the entire amine group leaving the benzyl (or more probably the tropylium) ion.

Secondary amines can be of two kinds, simple diaryl or alkyl aryl compounds.

(c) **ArNHAr′**

Compounds of this kind can usually be recognized from the molecular composition. Loss of one hydrogen is also likely. Most isomeric compounds will be primary amines which can be detected by the facile loss of hydrogen cyanide, a fission uncommon in the secondary amines.

(d) **$ArCH_2NHAr'$**

This class of amine will have a further very facile fragmentation corresponding to the formation of $ArCH_2^+$. Again any suspicion as to the category of the amine may be checked by treatment with nitrous acid.

(e) **ArNHR**

Mixed alkyl aryl amines will have the properties of the alkylamines previously discussed (Chapter 4, p. 77), as well as those of aryl secondary amines.

(f) $ArCH_2NHR$

This group will resemble those just discussed and will have in addition an abundant ion $ArCH_2^+$.

With tertiary amines the number of possibilities is greatly increased. Even so, the various categories will be merely a combination of all the properties listed above together with those of the alkylamines. Attempted nitrosation will reveal the category of the amine. The formation of the abundant ion of the form $ArCH_2^+$ will reveal the presence of this grouping and any extensive alkyl chain may be detected by the usual behaviour of these upon electron impact.

(g) ArCN, ArNC

Aromatic nitriles are particularly stable. They may often be recognized from the molecular formula, which will indicate the presence of a carbon in excess of the structural requirements of an aryl system, after allowing for the nitrogen. Thus the molecular formula $C_{13}H_9N$ requires for any stable structure the presence of a –CN or –NC grouping. The nitrile and isonitrile series can be distinguished by reduction, since the nitrile will be reduced to a primary while the isonitrile gives a secondary amine.

$$ArCN \xrightarrow{4H} ArCH_2NH_2 \xrightarrow{HNO_2} ArCH_2OH \longrightarrow \overset{+}{ArCH_2OH} + e$$

$$\overset{+}{ArCH_2OH} \downarrow \overset{+}{ArCH_2} + HO.\text{etc.}$$

$$ArNC \xrightarrow{4H} ArNHCH_3 \xrightarrow{HNO_2} ArN(NO)CH_3 \longrightarrow \overset{+}{ArN(NO)CH_3} + e$$

$$\overset{+}{ArN(NO)CH_3} \downarrow \overset{+}{ArNCH_3} + NO.\text{etc.}$$

Further treatment with nitrous acid will yield the appropriate derivatives.

5. TWO NITROGENS

Again two nitrogens may represent two different functions each containing nitrogen. Thus, an aromatic compound possessing an amine and a nitrile group would be included here. The various combinations that are possible can be classified by comparing the mass spectrum of the original substance with that obtained after reduction. If there is no change, the initial groupings were saturated and must be amines. If four hydrogens are taken up, the original molecule contained one nitrile or isonitrile and one amine or two amides etc. If eight hydrogens are absorbed, the original molecule contained two nitrile groups, two isonitriles or one of each.

The various categories may be further characterized by treatment with nitrous acid.

The determination of the molecular weight and the molecular formula after each experiment will be sufficient to reveal the hydrogen uptake (increase in molecular weight) and the effect of nitrous acid (molecular formula).

(a) ArN=NAr′

There remains a further important system which is rather stable under electron impact, the diazo compounds. This group is perhaps best recognized from the great stability of the parent molecular ion, the molecular formula and the colour of the sample. Diazo compounds contain a chromophoric group and many are highly coloured. Reduction will show an uptake of two hydrogen atoms and the production of a much less stable hydrazo compound.

(b) ArN=N—NHAr′

This group has received no systematic treatment and, apart from the probable loss of a single hydrogen atom and fission at the bond adjacent to the –NHAr′ group, nothing much may be predicted. The system is well known as tautomeric, although it is doubtful if, under the usual operating conditions of the mass spectrometer, both tautomers would arise should only one be introduced.

6. SULPHUR

(a) ArSH

The aromatic thiols are rather stable molecules, and the parent molecular ion is the base peak of the spectrum. There is some evidence for the loss of a single hydrogen, but the elimination of CS and HCS is nowhere marked, in contradistinction to the rather facile elimination of CO and HCO from phenols.

(b) $ArCH_2SH$

Benzyl thiols or analogues readily lose SH in the agreement with the facile fission at the β-bond.

(c) ArSAr′

These compounds are again very stable and the parent molecular ion is also the base peak. A moderate abundance of the ions Ar^+, $(Ar')^+$, ArS^+, and $Ar'S^+$ is observed.

(d) $ArCH_2SAr'$

These show the usual variation from the simple aromatic sulphides. The ions $ArCH_2^+$ and $Ar'S^+$ will be very abundant, in agreement with the preferred fragmentation at the bond joining them.

Mixed alkyl aryl sulphides will have the properties already ascribed to the alkyl and aryl sulphides with perhaps some extra fragments.

7. HALOGENS

(a) ArX

Aromatic halogen compounds of this type are very stable. There is little or no elimination of the halogen and, when this does occur, the aryl ion is formed in preference to X^+.

(b) $ArCH_2X$

In benzylic halides, on the other hand, the halogen is very easily eliminated.

In either case the molecular constitution is fairly easy to obtain.

F. Mixed Hetero-atoms

(a) ArNO

The aromatic nitroso compounds have not been extensively studied. However, they can be fairly easily reduced to the corresponding primary amine which, in turn, can be converted to the hydroxy compound. This, with a mass spectrum of each substance to determine molecular weight and composition, will demonstrate the nature of the starting material.

The spectra of the aryl nitroso compounds have an abundant parent molecular ion, some evidence for the loss of oxygen, rather more for the elimination of NO, and fragmentation of the aromatic system.

(b) $ArCH_2NO$

Compounds in this category readily lose NO which is sufficient to categorize the functional group and also to demonstrate that it is β-bonded to an aryl nucleus. For confirmation, if required, the compound can be reduced to the primary amine and thence to the alcohol in the usual way.

(c) ArONO

Aromatic nitrites may be expected to lose NO rather readily. The fact that a second oxygen remains together with a knowledge of the molecular weight are sufficient to avoid confusion with a nitroso compound ($ArCH_2NO$) in simple cases. The possibility of there being two functional groups, e.g. $ROArCH_2NO$, cannot, however, be ignored and while such a system will yield to detailed consideration of the cracking pattern, it is often simpler and more convincing to obtain the mass spectrum on reduction, followed by treatment with nitrous acid. If reduction yields a phenol, the original functional group was a nitrite. If a primary amine is obtained, the original molecule contained two separate functional groups, a nitroso group which reduced to the amine and a separate oxygen function.

(d) $ArCH_2ONO$

Such a compound will readily lose forty-six mass units from the parent molecular ion. This observation together with the fact that reduction will yield an alcohol should identify the functional group.

(e) $ArNO_2$

A very abundant parent molecular ion and much less abundant ions corresponding to the elimination of one oxygen, nitric oxide, and the nitro group are observed. This class may be readily distinguished from nitrites since reduction yields a primary amine.

(f) $ArCH_2NO_2$

The mass spectrum of the parent compound shows a ready loss of the nitro group. Reduction yields a primary amine which can be converted into the corresponding alcohol.

(g) ArCNO

Aryl cyanates may be detected by reduction to the primary amine with a loss of oxygen, and then conversion to an alcohol with removal of the nitrogen. All these compounds will have abundant parent molecular ions. The amine and alcohol will have the very abundant ion $ArCH_2^+$ in their spectra which may well be the base peak.

(h) ArNCO

The isocyanates upon reduction will lose oxygen to yield a secondary amine. This on treatment with nitrous acid will give an *N*-nitroso derivative. The determination of the molecular composition in each case will indicate that the original substance was an isocyanate.

(j) $ArCH_2CNO$

Cyanates of this type will readily lose forty-two mass units (CNO). Reduction will yield a primary amine $ArCH_2CH_2NH_2$, which will easily eliminate CH_2NH_2 from the molecular ion. There will, therefore, be an abundant ion at $(P-30)^+$. Treatment with nitrous acid gives the corresponding alcohol, the mass spectrum of which will be characterized by the very facile elimination of thirty-one mass units ($\cdot CH_2OH$).

(k) $ArCH_2NCO$

The parent molecular ion will lose forty-two mass units as with the cyanates above. Reduction yields a secondary amine and treatment with nitrous acid an *N*-nitroso derivative, the recognition of which will determine the original structure.

(l) ArAr′NOH

Oximes as a class may include diaryl or alkyl aryl and, if derived from aldehydes, aryl oximes. The melting points of these compounds are rather high and they are often difficult to introduce into a mass spectrometer except by special methods. It is much simpler to reduce them to the secondary amine and conduct the analysis on this.

Many other combinations of nitrogen and oxygen exist as mono-functional groups in aromatic chemistry.

(m) NNO

Two nitrogens and an oxygen may be present as the group $-N{=}\overset{\overset{\displaystyle O}{|}}{N}-$. It is clearly not possible to enumerate all of these. Accordingly, the basic procedure already mentioned should be applied, and a common sense interpretation of the results will determine the nature of the functional group.

For the present example, a mass spectrum of the original material will indicate a molecular weight (P^+) and a molecular formula which contains N_2O. Vigorous reduction will yield two molecular species each containing one nitrogen. The oxygen is lost from the system. If the molecular weights of the two species are added it will be found that their combined weights are twelve units less than the original. Exact mass measurements would show that the number of carbon atoms remains the same, although they were now distributed between two molecules. The cracking pattern of the products would confirm that each possessed a primary amine group, or this could be determined by treatment with nitrous acid to yield the corresponding phenols.

Since the molecule has split into two parts and lost an oxygen, each daughter molecule must have gained two hydrogens upon the reduction which cleaved the molecule.

$$\mathrm{ArN{=}\overset{\overset{\displaystyle O}{|}}{N}{-}Ar'} \xrightarrow{6H} \mathrm{ArNH_2 + Ar'NH_2 + [H_2O]}$$

The water formed does not appear in the molecular weights determined mass spectrometrically. The original formulation is thus obtained.

(n) Ar′ArSO

Aryl sulphoxides are stable under electron bombardment.

There is some evidence that fission occurs to yield both aryl ions (Ar^+, $(Ar')^+$). The ions $ArSO^+$ and $Ar'SO^+$ are also observed.

(o) $ArCH_2SOAr'$

In this category, the standard β-bond fission occurs to give the $ArCH_2^+$. The other ions just mentioned are also present.

(p) $Ar'ArSO_2$

Aromatic sulphones are relatively stable structures. The parent molecular ion is abundant. Fragmentation to yield Ar^+, $(Ar')^+$, $ArSO_2^+$ etc. is also observed. These compounds may be reduced to the sulphides, or even split to the aromatic thiol and the hydrocarbon. Moderate reduction to the sulphide will thus demonstrate that both oxygens were attached to the sulphur and will exclude the possibility that the compound is a phenolic sulphoxide.

Compounds of the form $ArCH_2SO_2Ar'$ will, as usual, show an abundant ion corresponding to $ArCH_2^+$.

(q) $ArSO_2$

Sulphinic acids or esters still have not been extensively studied. The acids may be expected to lose a hydroxyl and there will be some evidence for the aryl ion. The esters may be considered to fragment as follows.

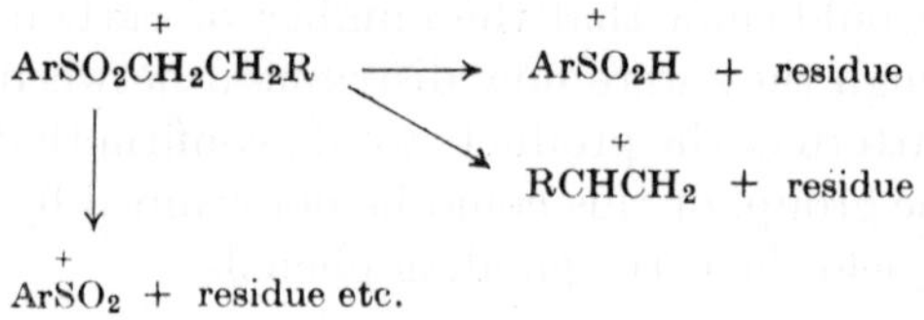

(r) **$ArCH_2SO_2$**

This series of acids or esters will have the same distinctive differences from the compounds immediately preceding, as have been discussed throughout this section.

(s) **$ArSO_3H$, $ArSO_3R$**

As far as is known, no sulphonic acids nor esters have been systematically examined. The relative involatility of the acid makes examination difficult. In the case of the esters, the normal fragmentations associated with these may be expected.

(t) **$ArSO_2X$, $ArSO_3X$**

Sulphinyl and sulphonyl halides seem to have been little studied. No doubt the parent molecular ions will prove to be abundant. Equally there should be a facile loss of the halogen.

(u) **$ArCH_2SO_2X$, $ArCH_2SO_3X$**

These halides may be expected to behave in a way similar to those above except that the molecule contains an extra methylene group. Fragmentation in this instance should occur between the carbon and sulphur bond, in addition to the elision of the halide.

G. Ausiastic Hetero-atoms

1. ALIPHATIC SYSTEMS

Hetero-atoms of this category are, as the name implies, incorporated in the carbon chain. Two such examples have already been reported, the methoxyl which was treated as a single entity and the secondary and tertiary amines which, for convenience, were discussed along with primary amines.

The problem in this class is that it is no longer possible by simple chemical operation to convert the functional group containing the hetero-atom to a primary hydroxyl or amine group. Moreover, since the presence of a double bond may affect profoundly the behaviour of the hetero-atom linkage, it is now desirable that reduction of the unsaturated centres be the main chemical adjunct to the normal mass spectrometric practice. This has the further advantage that, once the nature of the hetero-atom is known (a comparatively easy task provided that there is a reasonably abundant molecular ion), it is possible from the molecular formula of the reduced material to determine the number of rings present in the compound.

(a) One oxygen

The molecular formula on reduction of $C_nH_{2n+2}O$ indicates an acyclic ether. The problem of the location of the oxygen along the carbon chain is fairly readily decided from the information already available (see Chapter 4, p. 80). It is a matter of greater difficulty to determine the character of the two alkyl chains. If neither is too elaborately substituted this may be possible by a detailed analysis of the cracking pattern. As a final solution to the problem, it may be necessary to split the molecule with hydriodic acid, or some other suitable reagent, and examine the two products so formed.

$$R_2O + HI \rightarrow ROH + RI,$$

which would represent a fairly elaborate chemical method of analysis.

Should the molecular formula upon reduction correspond to $C_nH_{2n}O$, the structure must be cyclic. The common form is represented by tetrahydrofurans, the cracking patterns of which are fairly well established (see Chapter 5, p. 118). Monosubstituted compounds of this class may be readily identified by an extension of the usual method. Tetrahydropyrans have been little studied, but their breakdown would be very similar to the tetrahydrofurans. The same sort of fragmentation might also be expected for the cyclic structure $C_6H_{12}O$.

Systems containing two rings will have the formula $C_nH_{2n-2}O$. One advantage of reduction preceding analysis is, of course, that the presence of double bonds is automatically excluded. The fragmentations now involve those of an alicyclic as well as a heterocyclic ring. Again, the established pattern for this system has not been found, but there is little doubt that preferred cleavage will occur in the heterocyclic ring.

(b) Two oxygens

When two oxygen atoms are present, the system is most likely a monocyclic diether, or two cyclic mono-ethers either joined directly to each other or by a methylene chain. Oxides have been found occasionally in natural products. Lactones also belong in this classification. The relatively limited information about this group has already been detailed (Chapter 4, p. 87). Some chemical evidence may simplify the problem since, if the ring is opened and the acid group esterified, one obtains a hydroxy ester, which should be easier to identify. In the case of

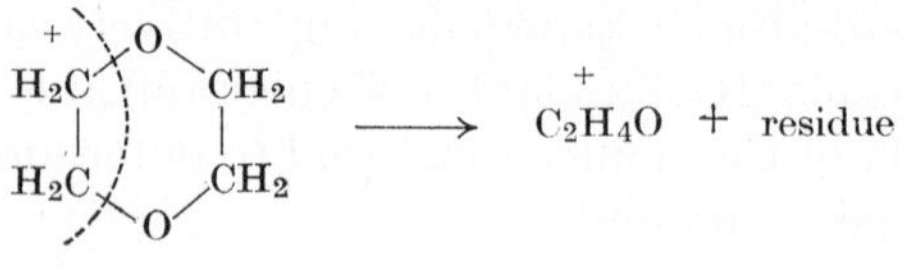

the cyclic diether, preferred fragmentation will again occur at the carbon, oxygen bond, as in the example of dioxane, when the main ion is formed by a double fission.

Substituents upon the carbon atoms may easily be recognized in the usual way.

(c) Three oxygens

Three oxygens in a saturated cyclic system is probably a polymeric structure formed from some aldehyde or, perhaps, a ketone. The material can almost certainly be depolymerized upon heating, and the original monomeric substance identified. Once the structure of this is known, the molecular weight of the polymer, if obtainable, will reveal the number of monomeric units in it and hence a structure may be obtained.

(d) One nitrogen

Molecules of the formula $C_nH_{2n+1}N$ can usually be prepared by the reduction of the partially unsaturated material. The commonest classes represented here are pyrrolidines and pyridines. Substituents may be of two kinds, on carbon or nitrogen. Compounds of the former class do not introduce any further complication. Fission of these compounds is favoured at the next but one bond to the nitrogen atom. The intermediate ion so produced is essentially the same as for an acyclic amine. The fragmentation of acyclic amines is well established, and the problem is essentially one of locating the position and nature of the carbon branch.

Chemically, the fully reduced cyclic systems which do not have a substituent upon the nitrogen are secondary amines. These may be detected by their basicity or, if necessary, by their reaction with nitrous acid when the *N*-nitrosamine is formed.

The *N*-substituted amines are analogous to the tertiary amines, which have been discussed earlier. The problems of fragmentation are well understood, and additional allowance must be made only for the carbocyclic ring.

Again, it is chemically possible to confirm the characteristic nature of the nitrogen even if only with a relatively unsatisfactory negative test; the tertiary nitrogen atoms are unaffected by treatment with nitrous acid.

(e) Two Nitrogens

Two nitrogens in a fully reduced system may either be in one ring when the formula will be $C_nH_{2n+2}N_2$ ($C_nH_{2n+1}N_2R$ if the cyclic diamine be substituted), or in two separate cyclic systems when the appropriate formula will be $C_nH_{2n}N_2$, $C_nH_{2n-1}N_2R$ etc. Again, chemical evidence is available to determine the nature of the nitrogen atoms. If both nitrogens are tertiary, the molecular weight of the compound will be the same after treatment with nitrous acid as before. Should the molecule contain one secondary and one tertiary nitrogen, treatment with nitrous acid will increase the weight by twenty-nine mass units as the new molecular formula will contain one nitrogen and one oxygen extra. Two secondary amines will yield a dinitroso compound. The new molecular weight will be fifty-eight units greater, and the molecular constitution will be augmented by N_2O_2.

Fragmentation of the cyclic systems will follow the usual pattern. The orientation of any substituent relative to the nitrogen atoms is, however, a matter of some difficulty except in the simpler compounds.

(f) Three nitrogens

Systems containing three aliphatic cyclic nitrogen systems are unusual, unless relatively large. However, some rather common compounds such as urotropine have this number. In addition to obtaining the molecular composition of the compound $C_4H_8N_3$, and demonstrating chemically that the nitrogens are all present as tertiary amines, the molecule may be decomposed on heating. The decomposition products will be easily recognized as ammonia and formaldehyde and, with this identification, the original molecule can be reconstructed.

(g) One sulphur

Acyclic sulphides are easily recognized by the general formula $C_nH_{2n+2}S$. Fragmentation of such compounds has already been reported (see Chapter 4, p. 85). The common modes of fission are at both the α- and the β-bonds. Recognition of the alkyl groups should not be so difficult in this instance, where the fragmentation is more extensive than in the acyclic ethers. Further, if necessary, the molecule may readily be cleaved at the hetero-atom to give the appropriate thiol and iodide.

Cyclic ethers and substituted cyclic ethers are of the general formula $C_nH_{2n}S$. Again, apart from the problem of ring size, which is not always readily determined, there is little difficulty in analysis. Moreover, if desired, these compounds may readily be cleaved to the acyclic ω-iodo alkyl alcohol before analysis.

The sulphur can be removed by treatment with Raney nickel and the resulting diolefin either analysed as such or reduced to the corresponding alkane. From the structure of the alkane, that of the original sulphide may be readily obtained.

(h) Two sulphurs

Several possibilities here exist. The compound may be of the formula $C_nH_{2n-2}S_2$ and belong to that class which contains two cyclic sulphide units. The two cyclic structures may be separately attached to a further aliphatic chain, or to each other. If the former situation exists, the problem is not too difficult. The molecular ion will cleave preferably at the points where the rings are attached and unless there is a second substituent upon the ring, ring size can then be determined. Again, in complex systems, the compound may be treated to remove the sulphur and hydrogenated to the appropriate alkanes, the alkanes identified and the original molecule reconstructed. Such a method has several advantages.

It is possible to obtain a sulphide in which the ring structure is carbocyclic and the sulphur is contained in an acyclic substituent of the rings. This series is isomeric with the alkyl substituted cyclic sulphides. Their behaviour upon chemical treatment, however, is widely different. Treatment of the heterocyclic system with Raney nickel will result in the formation of the tetraolefin, which can then be hydrogenated to the corresponding alkane. Thus, if the parent ion of the initial compound is P^+, it falls to $(P-68)^+$ after treatment with nickel and rises to $(P-60)^+$ on reduction. The structure of the alkane is then determined.

If, on the other hand, the molecule is of the form RSS_2R', where R′ comprises the main carbon skeleton, treatment with nickel will yield at least two products resulting from the elimination of the sulphurs. One, or two of these, containing the alicyclic system, will yield alkanes of the formula C_nH_{2n} upon reduction, the others, alkanes of the formula C_nH_{2n+2}. These may be identified in the usual way. Certain precautions must be observed. It is important to determine the molecular weights and compositions of the alkanes to ensure that all the carbon atoms present in the original sulphide are accounted for. Also, in order to decide which of the possible positions in the alkane was originally attached to the sulphur atom, the double bonds in the alkenes must be located. Finally, in order to remove the remaining ambiguity as to which end of the double bond was so bonded, an analysis of the mass spectrum of the original sulphide aided by the known skeletal structure is undertaken. The two sulphur atoms may be incorporated in one ring when the determined structure will be $C_nH_{2n}S_2$. The two hetero-atoms may be remote from each other, as in the thio analogue of dioxane, when a cracking pattern similar to the last-named will be obtained, or they may be together in the form of a disulphide. This can be detected by treatment with Raney nickel. If the compound is a disulphide, the main carbon chain in a cyclic compound will remain unbroken. If, however, it is two distinct sulphides then two hydrocarbon chains will be obtained.

H. Mixed Hetero-atoms

When two hetero-atoms are present in an ausiastic system, they may be attached either to one another, independent but near neighbours, or remote from each other. The systems that are considered remote from each other, i.e. separated by three or more carbon atoms, are the simplest to deal with, as the influence of each of the hetero-atoms is exerted independently. When the two atoms are nearer to each other than this, the cracking pattern caused by the one will interfere with that of the other, and the two patterns will be modified. Finally, when the two atoms are contiguous, the system may well have characteristics which derive mainly from them collectively rather than from the independent influence of each.

The combinations possible for two atoms are limited and these are discussed in some detail. Obviously as the number of hetero-atoms increases so do the possibilities, and a general classification is impossible.

1. NITROGEN AND OXYGEN

An acyclic system containing a secondary amine and an ether linkage may be difficult to analyse. Provided that the aliphatic system containing them is fully reduced, it is possible to avoid confusing such structures with nitrites and acid amides. The formula for such a compound will be $C_nH_{2n+1}NO$. Probably the simplest method of proceeding would be to obtain the cracking pattern of the original and then attempt to cleave the ether linkage with hydriodic acid. The resulting iodide and alcohol may be analysed, and the composition of each determined. One of these will not contain the nitrogen atom and, since it will be a simple alcohol or halide, analysis of this moiety should prove possible.

The likelihood does exist that fission will not occur exclusively on one or other side of the oxygen. In these instances, four products will be obtained. The mixture must either first be separated by a gas–liquid chromatogram, or the mass spectrum of the mixture examined and the composite cracking pattern subsequently resolved.

If the molecular formula proves to be of the form $C_nH_{2n-1}NO$, then the system contains either a heterocyclic or an alicyclic system. Should the ring be carbocyclic, it may be recognized from the cracking pattern, since cleavage at the point of substitution on the ring often occurs preferentially. Provided that this fragment is recognized, the size of the ring may be deduced.

For the positioning of the nitrogen and oxygen functions one can proceed further by the methods already outlined.

A further possibility is that one of the hetero-atoms occurs in a cyclic and the other in an acyclic structure. Treatment with hydriodic acid would decide the arrangement. Should the nitrogen be in the cyclic system, then the most probable result of this treatment is to cleave the bond at the oxygen atom. If two products are formed, it is possible that both may be analysed, if four, then either separation followed by analysis, or a mixture analysis must be undertaken.

Finally the two hetero-atoms can be contained in one ring. Fragmentation may be expected at the bonds β to the oxygen. This will be marked if the oxygen lies between methylene groups. Similar fissions will be apparent for the nitrogen also. In the case of alkyl substituted heterocycles which contain both nitrogen and oxygen, some chemical assistance may be needed to simplify the analysis; nitrous acid should be used to identify the character of the nitrogen.

As an example, the analysis of a simple compound of this class follows. The molecular ion is quite prominent, being 25·3% of the base peak ($m/e = 43$). The molecular weight of the parent ion is 101, and precise mass measurement indicates a molecular formula of $C_5H_{11}NO$. There is a reasonably abundant ion $m/e = 43$ (the base peak) which is consistent with the partial structures C_2H_5N and C_2H_3O. Chemical evidence shows that the nitrogen is tertiary. Among the ions found is $m/e = 71$, corresponding to the loss of CH_3NH from the molecular structure. Bearing in mind the molecular formula and the fact that there are no double bonds present, the structure is cyclic and contains one ring. If the parts already identified are considered, there are two carbon atoms to place. There is evidence in the spectrum of a moderately abundant ion $m/e = 27$ corresponding to the formation of the vinyl ion ($C_2H_3^+$). Further, the abundance of the methyl group (19%) is large enough to indicate that the structure almost certainly contains a methyl group as such and, moreover, it is attached in a way that makes cleavage very likely. The material is N-methylmorpholine (Chapter 4, p. 92).

2. SULPHUR

Compounds which contain sulphur are more amenable to analysis. Firstly, the identification of the sulphur atom is somewhat simpler even with single-focusing instruments since the sulphur isotopes are readily recognizable. Secondly, if it is necessary to remove the sulphur to aid in analysis, this may be readily effected by the use of Raney nickel or even hydrogenolysis (Emmett, 1957). The resulting olefinic compound may then be reduced to the corresponding alkane or alkanes and these identified.

J. Aromatic Type Compounds

This category contains many common compounds which include

pyridine, furan and thiophene, and also these fused to a benzene ring, indole etc. The analytical problems posed are many and varied and while it is not possible to analyse all the systems, some general principles will be given as a guide.

The chief distinction in this series is that, while with the heterocyclic systems so far considered it was possible to reduce the molecule to the saturated analogue without difficulty, this is not practicable for the present series. On the other hand, the very stability which makes reduction of an indole a difficult process means, in turn, that it is likely to be a prominent ion (if not actually the base peak) even if it has a quite complex ring system attached. Further, there is the considerable difficulty, if not near impossibility, of determining the orientation of substituents solely upon the evidence contained in the fragmentation pattern.

1. ONE OXYGEN

The presence of one oxygen does not really create a serious problem, since in nearly all systems it is possible to reduce it to a saturated cyclic ether which may then be analysed by the methods already reviewed.

Certain exceptions to this general principle exist, however. One such is dibenzofuran, in which the double bonds of the furan group are part of the aromatic rings. These compounds are not easily analysed, since the parent molecular ion is also the base peak, and the rest of the cracking pattern, with the exception of the $(P-1)^+$ ion, is slight; similar difficulties would be encountered in the analogous dibenzopyran.

If the system could be split by reductive fission the analytic problem would certainly be eased, but not solved. Reductive fission of dibenzofuran would lead to a monohydroxydiphenyl. The formula of $C_{12}H_{12}O$ would be sufficient to characterize this molecule. A diphenyl structure is most likely and would in this instance be recognized by the presence of the fragment ion $m/e = 77$. The phenolic nature of the hydroxyl group is readily confirmed by the loss of carbon monoxide from the parent molecular ion.

$$\overset{+}{C_{12}H_{10}O} \rightarrow \overset{+}{C_{11}H_{10}} + CO$$

Not all cases are so easy, but the general principles are unaffected by the presence of additional aromatic rings or simple alkyl substituents.

The alternative problem, which may be discussed for the sake of completeness, namely that the cyclic structure is purely carbocyclic and the ether linkage is in the aliphatic part of the molecule, is more simply analysed. Should the aromatic system be directly attached to a carbon of the aliphatic portion, one may expect the characteristic abundant fragments. If the group be phenyl, the ion $m/e = 91$ (corresponding to the tropylium ion) will be abundant, while naphthyl will yield $m/e = 141$. As before, the orientation of substitution on the naphthalene nucleus is difficult without other evidence.

2. TWO OXYGENS

The analysis of systems containing two ether oxygens follows exactly the same pattern. There may now be one oxygen in a benzofuran system and one in an alicyclic group. The alicyclic structure can be analysed by the methods given above, and the aromatic moiety obtained by inference.

One further possibility which must be discussed is that of the aryl-oxy system. Here the main fragment ion has the formula $C_6H_5OH^+$ and corresponds to β-bond fission with hydrogen rearrangement. Similar fragmentation in the naphthalene derivatives yields the abundant ion $m/e = 143$ ($C_{10}H_7O^+$).

3. ONE NITROGEN

Unlike the conjugated oxygen-containing derivatives, those containing nitrogen, such as pyridine, quinoline, and indole, are not readily

TABLE XXVII

Compound	Base peak m/e	Base peak formula	Fragment elided
2-Methylpyrrole	80	C_5H_6N	H
N-n-Butylpyrrole	80	C_5H_6N	C_3H_7
2,4-Dimethylpyrrole	94	C_6H_8N	H
3,5-Dimethylpyrrole	94	C_6H_8N	H
Indole	117	C_8H_7N	—
2-Methylindole	130	C_9H_8N	H
3-Methylindole	130	C_9H_8N	H
5-Methylindole	130	C_9H_8N	H
6-Methylindole	130	C_9H_8N	H
1,3-Dimethylindole	144	$C_{10}H_{10}N$	H
2,3-Dimethylindole	144	$C_{10}H_{10}N$	H
2,5-Dimethylindole	144	$C_{10}H_{10}N$	H
2,6-Dimethylindole	144	$C_{10}H_{10}N$	H
3-Ethylpyridine	92	C_6H_6N	CH_3
2-Methylpyridine	93	C_6H_7N	—
6-Methylquinoline	143	$C_{10}H_9N$	—
7-Methylquinoline	143	$C_{10}H_9N$	—
2,6-Dimethylquinoline	156	$C_{11}H_{10}N$	H
10,10-Dimethylacridane	194	$C_{14}H_{12}N$	CH_3

reduced. Therefore, the heterocyclic system must be identified if possible without chemical assistance. Fortunately, many of the ions formed by these compounds or their alkylated homologues are abundant and are recognizable structurally. The base peak obtained from a series of these compounds is shown in Table XXVII. Once the functional system has been obtained, analysis, apart from orientation problems which are always difficult, is fairly simple. In compounds containing one nitrogen in the heterocycle, the remaining problem is that of assigning a structure to the residual carbons and hydrogens.

Should the hetero-atom be in the chain and not in the cyclic structure, the base peak will arise from the aryl group present. The nature of the nitrogen atom can then be determined by treatment with nitrous acid in the usual way, and the analysis continued upon the product so obtained.

4. TWO NITROGENS

The combinations that may be encountered here are covered in the preceding analyses by a combination of the possibilities there listed taken separately. Again, it is possible that the two nitrogens are present in one ring. Compounds of this class include the "azines".

Such compounds, in general, can be reduced at least partially and, if possible, it is convenient to reduce them to the saturated cyclic compound. Treatment with nitrous acid will reveal the character of the nitrogen bonding and an analysis of the cracking pattern should then be possible.

5. ONE SULPHUR

Sulphur-containing substances are much more amenable to treatment than either oxygen or nitrogen compounds. This arises from the observation that sulphur may be removed even from aromatic type compounds with Raney nickel. The remaining hydrocarbon may then be hydrogenated and identified.

If the compound is simply a sulphur-containing heterocyclic, extraction of the sulphur followed by reduction (or hydrogenolysis) and analysis is all that is necessary. The alkane may be identified, which will assist in positioning the double bonds in the intermediate product. Final formulation of the original structure will depend upon an assessment of the initial cracking pattern.

6. TWO SULPHURS

Should an aromatic structure contain two sulphur atoms, these may be treated similarly. An initial mass spectrum will yield the molecular weight and formula. Removal of the two sulphur atoms gives one or more hydrocarbons. The molecular weights will show whether the system was an aromatic disulphide or not. Reduction to the alkane or, if an intact aromatic group still remains, to an arylalkane should then be carried out, and a mass spectrum of the product obtained. From this pattern, the molecular structure of the reduced material can be determined and after working back through the compounds, the original structure could be deduced.

The possibility that an aromatic ring will remain intact should always be considered, since compounds such as benzothiophene will lose the sulphur, but the aryl group will not be altered under the conditions that are normally employed for the reduction of aliphatic double bonds.

K. Mixed Hetero-atoms

In the aromatic field, an assemblage of hetero-atoms is rather less important than in aliphatic or cyclic systems. Commonly, the heterocyclic ring, which contains one or both or all atoms, is not easily ruptured, and the cyclic system remains intact. This makes it particularly easy with a double-focusing instrument (where the exact mass of fragment ions may be measured) to determine the molecular constitution of the ring and hence obtain some evidence as to the overall structure. Again, it is a rather difficult matter to determine the orientation of one hetero-atom with respect to another.

Should one of the hetero-atoms prove to be sulphur, the problem is somewhat simpler, since this atom at least can be removed and the analysis conducted in the usual way with one fewer hetero-atom present. Following the usual arguments the original molecule may be reconstructed.

L. Mixed Systems

Finally, some discussion should be given to systems in which both epeisactic and ausiastic hetero-atoms are present. Obviously, such a combination covers a very large field of organic chemistry, and it is only practical to give a general method of analysis.

The molecular weight, molecular formula, and cracking pattern of the material ought first to be obtained. The molecular formula of the compound will reveal the nature of the hetero-atoms present, if any. The abundance of the parent molecular ion is also instructive. Should the parent ion appear weak in relation to the molecular weight, a decision which unfortunately comes with experience rather than in accordance with precisely stated rules, the molecular structure is either highly branched or contains a hetero-atom in the main carbon chain. As has been previously discussed, this has the effect of lowering the abundance of the parent ion unless the hetero-atom be sulphur (see Chapter 2, p. 40).

The mass and composition of the base peak is often very informative also. The occurrence of such a peak at $m/e = 43$ ($C_3H_7^+$) is usually good evidence for the presence of a propyl or isopropyl group, $m/e = 57$ due to ($C_2H_3NO^+$) has in all the cases where we have encountered it indicated a cyanate attached by a phenolic group to an alkaloid residue; equally $m/e = 57$ ($C_4H_9^+$), recognized as such by its exact mass, is often associated with a n-alkyl chain of some length, $m/e = 91$ ($C_7H_7^+$) represents a benzyl structure present in the original molecule, $m/e = 93$ (C_6H_5O) is characteristic of a phenoxyl group, and $m/e = 93$ ($C_7H_9^+$) is the base peak of certain terpenes such as α-pinene.

TABLE XXVIII

m/e	Species	Precise mass	Probable environment of the fragment*
15	$\cdot CH_3$	15·0235	*gem*-Dimethyl, allylic methyl
17	$\cdot OH$	17·0027	Hydroxyl unable to eliminate as water
27	$\cdot C_2H_3$	27·0235	Often derives from an ethyl group
29	$\cdot CHO$	29·0027	Terminal aldehyde group
	$\cdot C_2H_5$	29·0391	As secondary or tertiary group
39	$\cdot C_3H_3$	39·0235	An aliphatic three-carbon fragment
43	$\cdot C_3H_7$	43·0548	Propyl or isopropyl group
55	$\cdot C_3H_3O$	55·0184	Ether or similar group
	$\cdot C_4H_7$	55·0548	Four-carbon chain
57	$\cdot C_2H_3NO$	57·0215	Cyanate
77	$\cdot C_6H_5$	77·0391	Unsubstituted phenyl group
91	$\cdot C_7H_7$	91·0548	Benzyl or tropylium group
93	$C_6H_5O\cdot$	93·0340	Phenoxy group

* The environment of the group may sometimes differ considerably from that suggested. These remarks are intended as a general guide only.

Reduction of the original material will reveal the number of reducible bonds on redetermining the molecular weight. If oxygen is present, the differences between the spectrum of the original and reduced species will indicate whether or not the nature of the linkage has changed. An aldehyde will now appear as a primary and a ketone as a secondary alcohol; ether linkages will not be affected.

Should nitrogen have been detected, it is useful to try the effect of nitrous acid before and after reduction and then compare the result. Some nitrogen compounds, such as imines or anils, are not likely to react with nitrous acid before reduction. Afterwards, however, they will yield alcohols which can, by the nature of the fragmentation pattern obtained, be shown to be primary.

As already stated, the aromatic part of a structure, by fission at the β-bond, often cleaves completely from the carbon chains to which it is attached. Exact mass measurement will yield the composition of the group removed.

CHAPTER 8

Logical Argument

The complete, or partial, analysis of a mass spectrum depends largely upon logical arguments of admission, or exclusion, of a given hypothesis. Even in rather simple examples, the development and investigation of such a hypothesis may be long and difficult. Accordingly some attempt is here made to set out the problem in terms of Boolean algebra, which is very suitable for the purpose. While it is true that methods of trial and error will produce the same result, the systematic treatment does ensure that no combination of possibilities, or some less likely contingency, is overlooked. The algebra used here represents an elementary section of the algebra of propositions, a subject until recently rather neglected, and a short summary of the axioms may usefully be given.

A letter is used to denote a class of elements that possess a certain property. Let this letter be p and let q represent a class with a further property.

Then $p+q$ is used to denote a class the members of which are either members of p, or of q or of both. The symbol pq represents individuals which are members of both p and q.

From these definitions, the following statements follow

$$p+q = q+p, \quad pq = qp$$

which is the commutative law of algebra,

$$p+(q+r) = (p+q)+r, \quad (pq)r = p(qr)$$

the associative law,

and
$$p(q+r) = pq+pr$$

Further, we introduce a symbol 0 to represent a null class (a class with no members) and a symbol 1 for a universal class which contains all; they are not those of ordinary arithmetic, but they obey the same rules.

If we further define a class p' which does not possess the property of class p then

$$p+p' = 1 \quad \text{and} \quad pp' = 0$$

A few more formal relationships and the list of the algebra, as far as is now needed, is complete.

$$(p')' = p$$

$$p\times 0 = 0; \quad p\times 1 = p; \quad 1' = 0$$

$$p+0 = p; \quad p+1 = 1; \quad 0' = 1$$

Further relationships which are not required in the present discussion may be found in standard works on the theory of sets (Whitesitt, 1961). Moreover, in order to keep the argument at an intuitive level in the absence of formal proof, several examples of varying complexity are given.

Consider an alcohol of which the skeleton, shown by the analysis of the parent hydrocarbon, is that of n-pentane. The problem is to place the hydroxyl group.

Number the carbon atoms 1–5, inclusive. Suppose the hydroxyl is attached to the first carbon and this is represented as p_1. Then the negation is p_1'. A similar representation may be employed for all five carbons.

Now $p_1+p_1' = 1$ as also does p_2+p_2' by the formulas already given. Multiplying all five such equations gives

$$(p_1+p_1')(p_2+p_2')\ldots(p_5+p_5') = 1$$

and, expanding and simplifying this expression, one gets

$$p_1p_2'p_3'p_4'p_5'+p_1'p_2p_3'p_4'p_5'+\text{etc.} = 1$$

The terms which are obviously zero have been eliminated. These include $p_1\,p_2\,p_3'\,p_4'\,p_5'$ etc. There is only one hydroxyl group and if it is on p_1 then it cannot be on p_2. Therefore any product containing $p_1\,p_2\,p_3'$ must be zero. Since this equation equals unity, it must contain the correct proposition. The equation may be obtained in other ways and can even be set down by inspection following the pattern given.

An examination of the fragmentation pattern provides the solution. There is no abundant ion $m/e = 31$, nor $(P-31)^+$, which is consistent with the view that the hydroxyl is not terminal. Hence, the products incorporating p_1 and p_5 are zero and can be eliminated. The ion $m/e = 45$ is abundant and represents the ion $C_2H_6O^+$. Therefore p_2 and p_5 equal unity, since it is a true proposition and the products containing them are equal to one.

$$p_1'\,p_2\,p_3'\,p_4'\,p_5' = 1 = p_1'\,p_2'\,p_3'\,p_4\,p_5$$

These are obviously equivalent positions in the present problem and the material is pentan-2-ol or s-amyl alcohol.

The formal logic is more cumbersome than intuitive deduction in this instance because a very simple illustrative example was used.

A more troublesome problem arises when there is a larger skeleton and two functional groups which are known not to be on the same carbon. Two algebraic equations are now required and the correct proposition must be contained in each one. This type of analysis may be set out formally for m substituents upon n carbon atoms with the limitation that no two substituents occur upon the same carbon. A set of products is obtained as before. By designating the first property p, assigning numbers for the carbon atoms and denoting the negation of the property by p',

$$P = \sum_{1}^{n} (p_1 p_2' \ldots p_n') = 1$$

Similarly for the second property of

$$Q = \sum_{1}^{n} (q_1 q_2' \ldots q_n') = 1$$

and so on.

Then the correct combination of propositions is found somewhere in the product

$$PQR \ldots = 1$$

As a further example, consider the spectrum of 2-methyl-2,4-pentanediol. There are two positionally exclusive functional groups, so that $m = 2$ and the solution is of the form $PQ = 1$.

The carbon skeleton may be determined by reducing the compound to the hydrocarbon, which would then be recognized as isohexane. This skeletal arrangement may be numbered in the following way

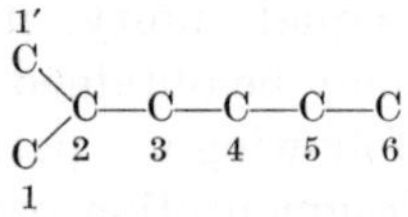

The problem now is to position the two hydroxyl groups along the chain. The base peak of the spectrum is at $m/e = 59$. This ion must have the formula of C_3H_7O or, less likely, $C_2H_3O_2$, which implies that either one oxygen is associated with a three-carbon fragment, or that both oxygens are on adjacent carbons. Exact mass measurement indicates that the former interpretation is correct. The ion $m/e = 45$ (24%) is one of the few remaining abundant ions and this too must contain an oxygen atom being of the form $C_2H_5O^+$. There is no abundant ion at $(P-31)^+$ or 31^+, which are so often characteristic of terminal hydroxyl groups and hence the ion 45^+ must have the formula CH_3CHOH. These two carbon atoms are five and six on the diagram, the only two that fulfil the required condition. The function Q has been evaluated for $p_5 = 1$. Again, since there is little or no evidence for a $-CH_2OH$ group, the other hydroxyl is on carbon number 2, the most reasonable way of reaching the formula for $m/e = 59$. The function P is evaluated, since $p_2 = 1$. The formula becomes

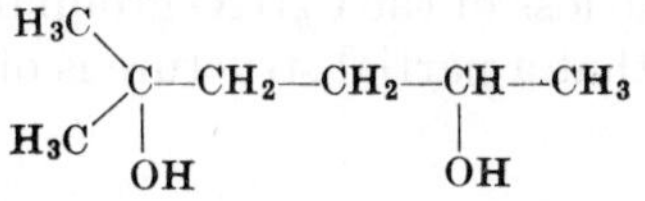

The systematic approach to analysis may be extended to larger molecules. The main difficulty remains a convenient method of enumerating the hydrocarbon.

The mathematical symbolism is not used explicitly in the following examples, although the method is evident.

A similarly complicated structure is that shown by pregnanolone. The

TABLE XXIX

m/e	Consolidated GEC M.S. 21–103	A.E.I. M.S.9	m/e	Consolidated GEC M.S. 21–103	A.E.I. M.S.9
67	95·1	85·1	121	49·4	53·0
71	62·5	47·1	133	36·8	—
77	51·3	—	135	34·6	—
79	95·1	83·8	147	41·6	45·0
80	100	100	161	34·0	—
84	75·1	100	215	80·9	92·2
91	77·3	47·1	230	29·0	33·8
93	89·0	82·6	242	22·9	29·9
94	40·9	45·0	246	20·3	—
95	87·1	96·3	257	23·0	22·1
105	60·8	68·4	285	30·8	32·5
107	79·3	80·1	300	89·0	15·1
108	38·3	43·4	301	23·3	30·9
109	40·7	44·2	318(p)	12·9	22·1
119	38·9	43·0			

more abundant ions from this compound, as recorded upon two different instruments, are listed in Table XXIX.

The precise mass measurements made with the double-focusing mass spectrometer indicate that the parent molecular ion has the formula

$C_{21}H_{34}O_2$, while the fragment ions $m/e = 84$ and 215 (101·3% and 92·2%) have the formulae C_5H_8O and $C_{16}H_{23}$. Since the overall fragmentation pattern of the corresponding alkane indicates a skeletal structure of a steroid, the problem reduces to that of positioning the two functional groups.

The ion $m/e = 215$ represents the loss of the side chain and ring D, and corresponds to the loss of the C_5H_8O group together with molecular water. This suggests that a partial structure is of the form

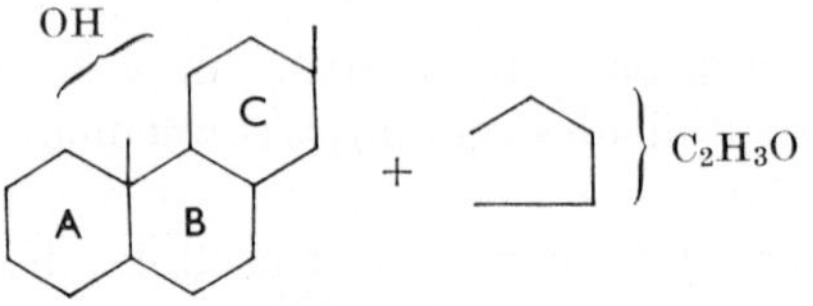

and that the hydroxyl group is not contained on ring D.

One of the common fragmentations in the steroidal skeleton is the loss of ring A in accordance with the fragmentation

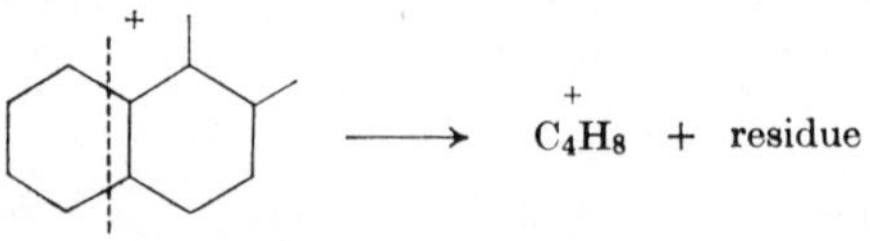

In the spectrum obtained, there is no ion corresponding to the loss of fifty-six mass units, nor indeed an abundant ion in the region, since one must allow for a single hydrogen transfer in either direction, of $m/e = 55$, 56, or 57. There is, however, an abundant ion $m/e = 71$ (62·5%) which has been shown to have the composition $C_4H_7O^+$ by exact mass measurement. This locates one oxygen on ring A which, since it is already known that a hydroxyl group is present on ring A, B, or C, identifies the nature of the group.

The molecular formula ($C_{21}H_{34}O_2$) requires five double-bond equivalents of which four are accounted for in the tetracyclic structure. The remaining double bond equivalent must be lost in the fission which removes the side chain and ring D since (excluding the molecular water eliminated from ring A) the composition of the material lost is C_5H_8O having two double bond equivalents only one of which is accounted for by ring D. There is a moderately abundant ion $m/e = 257$ (23%) having the composition $C_{19}H_{29}^+$, which corresponds to the loss of molecular water and a CH_3CO group. Accordingly, the side chain must be of that form and the complete structure is

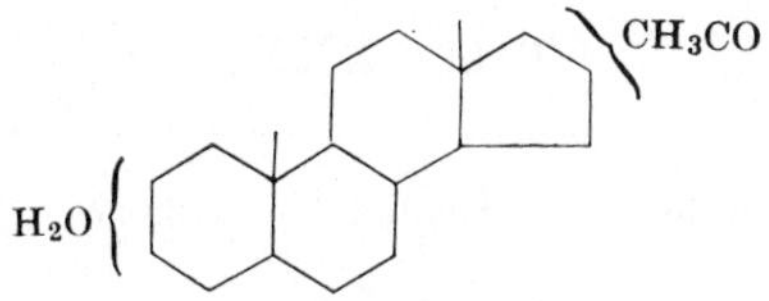

It should now be clear that, while this technique is of assistance in enumerating and evaluating the possibilities, there are nonetheless many conditions in which the simple analysis so far detailed is insufficient to locate the substituents. Additional information is therefore needed. This can often be obtained in the form of the so-called correlation studies previously mentioned (see Chapters 3, 4 and 5) where the spectra of the substance, the alkene, alkane and possibly other closely related compounds can be compared. The alkane is particularly valuable as a comparison, since it enables one to determine the skeletal arrangement of the molecule, as in the following example. Comparison is here made between the original compound, a dihydroxy-acid, a mono-alcohol, an alkene and an alkane. The abundant ions in the cracking-pattern of the alkane shows it to be a steroid. The parent molecular ion $m/e = 330$ indicates a formula of $C_{24}H_{42}$ which, since we know that the molecule does not contain a double-bond, represents a tetracyclic steroid. The abundant ion $m/e = 217$ results from the loss of the side chain together with ring D. Hence the side chain contains five carbon atoms.

Since occasionally (as in lophane) compounds are known which have a single methyl group at carbon 4, it is desirable to confirm the structure of ring A. The known fragmentations indicate that the ion corresponding to

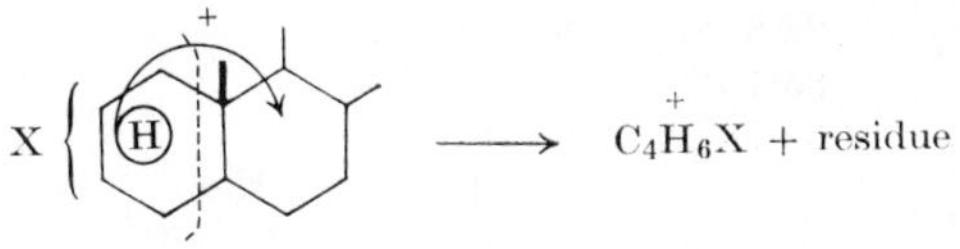

is very abundant. The abundant ion in the alkane is at $m/e = 55$ whence $X = H$ and the structure of the alkane is

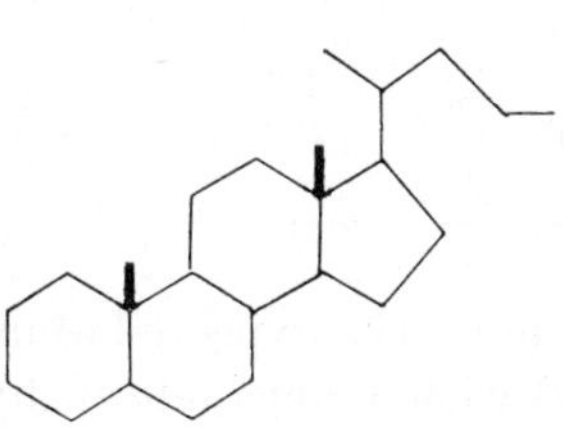

By analogous arguments, the hydroxyl group in the alcohol is contained neither on the side chain nor in rings A and D. An examination of the

TABLE XXX

m/e	Acid	Alcohol	Alkene	Alkane
28		17·9		
29		15·8		
32			16·2	
41		41·6	19·0	43·1
43		37·6		45·4
55		55·0		66·3
65	23·9			
67	58·3	32·4		41·1
68				52·5
69	24·4	24·6		25·0
77	59·8			
79	80·6			25·0
81	58·9	44·8	14·1	48·5
83				15·1
91			12·6	
93				24·2
95		33·4		49·0
97				12·0
98				94·6
105	79·5		23·0	14·8
106			19·2	
232				11·5
234		13·6		
255	60·9			
257		10·0	10·2	13·0
268	22·8			
282	83·1			
283	35·7			
313			13·6	
315				30·9
328		19·6	15·0	
330				42·1
338	50·4			
339	14·4			
346		1·2		
356	10·6			

spectrum of the alkane shows the expected abundant ion corresponding to cleavage across ring B with a concomitant hydrogen migration.

The corresponding migration accompanying ring cleavage is also observed in the alcohol. This proves that the oxygen is not attached on carbon 6 and by analogy with the fission of the alkane is probably not

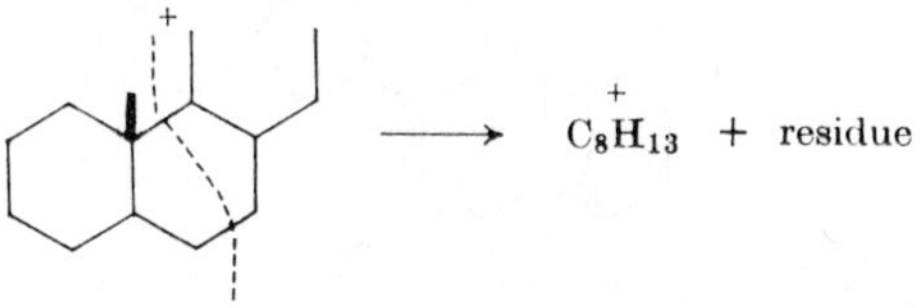

on 7. There now remain only two possible positions, namely 11 and 12. Fragmentation in the alkane yields an ion $m/e = 151$ ($C_{11}H_{19}^+$) which occurs neither in the alkene nor the alcohol. The positioning of the double bond at 11 and 12 is consistent with the observed spectrum. The presence of this bond would minimize such a cleavage for the alkane in favour of the following:

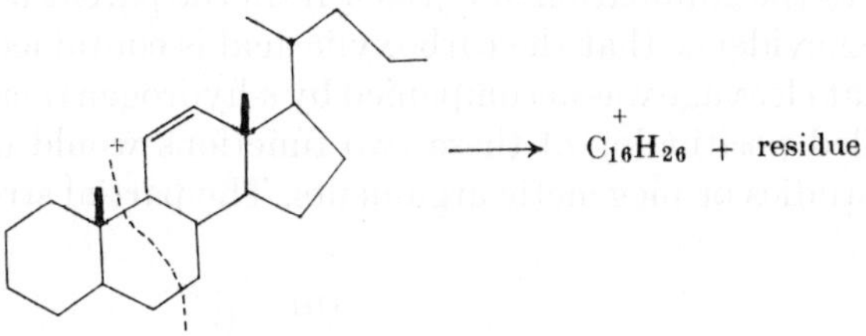

which effectively places the hydroxyl at one or other of these positions. The most abundant ion in the alcohol, which is relatively insignificant in the hydrocarbons, occurs at $m/e = 257$ and corresponds to the loss of a C_5H_{11} radical plus water. This grouping must represent the entire side-chain together with the hydroxyl. Steric interactions between the side-chain and neighbouring atoms must be the same for the adjacent methyl in all these compounds. On the other hand, it is well known that 1,3-interactions are particularly important in steric effects (Newman, 1956), and this suggests that the group that gives rise to the new steric factor is present upon carbon-12.

The acid shows many differences from the alkane, alkene and alcohol, and the analysis of the dihydroxy acid proceeds as follows. Since it is related to the alcohol whose structure has been determined, one hydroxyl is positioned at carbon 12. The compound possesses a steroid skeleton and there will be abundant ions corresponding to the following fissions with concomitant hydrogen migration.

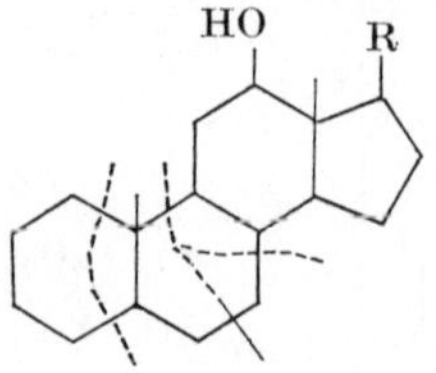

Ions of formulae $C_8H_{11}^+$ and $C_9H_{13}^+$ are present in the spectrum which indicate that there is a double bond present in each of these ions. Since the carbon skeleton was known to be saturated in the parent molecular ion, it follows that, in addition to the cleavages shown, elimination has also occurred. Therefore a hydroxyl group must have been present on ring A. The structure of rings A and B is now partly determined.

It was observed in the mass spectrum of the alcohol that the presence of a hydroxyl group at carbon 12 resulted in the facile cleavage of the side-chain at carbon 14. The most abundant ion at high mass $m/e = 282$ corresponds to the elimination of $C_5H_8O_2$ from the parent molecular ion. This is strong evidence that the carboxylic acid is contained in the side-chain and that cleavage was accompanied by a hydrogen re-arrangement.

The detailed positioning of these two functions would depend upon correlation studies or biogenetic arguments. The partial structure is

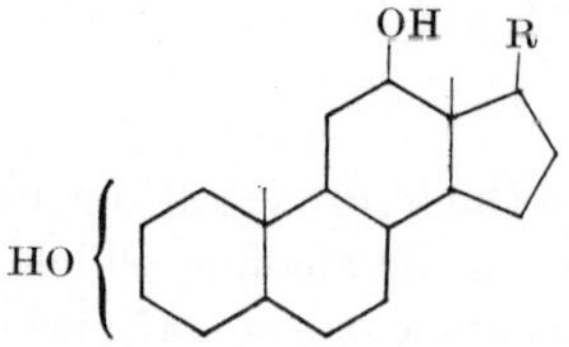

On biogenetic arguments the hydroxyl would most probably be at position 3. The carboxylic acid must be a terminal group, whence the structure of the side chain is either

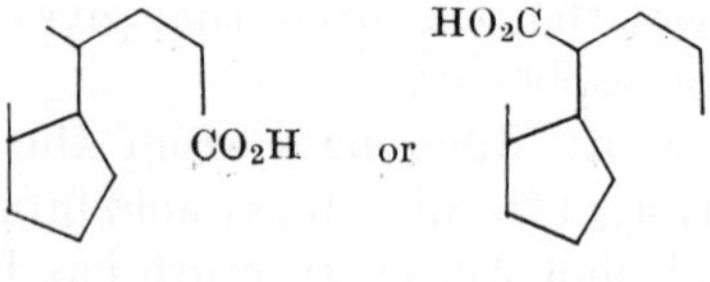

The former is the correct choice.

The recognition of a phenyl group in an organic compound is often of very great importance in analysis. Many naturally occurring substances contain either a substituted or unsubstituted phenyl. A previous investigation of aromatic hydrocarbons indicated that these structures tend to eliminate acetylenic units, and such cleavages may be used to deduce the structure of the original polycyclic hydrocarbon (Reed, 1960). Elimination has also been observed in alkylated benzene systems, as

TABLE XXXI

Compound	Base peak	(Base peak − 26) %
Benzene	78	17·65
Toluene	91	10·76
Ethylbenzene	91	8·53
1,2-Dimethylbenzene	91	6·35
1,3-Dimethylbenzene	91	6·49
1,4-Dimethylbenzene	91	6·02
n-Propylbenzene	91	8·73
Isopropylbenzene	91	1·84
1-Ethyl-2-methylbenzene	105	7·51
1-Ethyl-3-methylbenzene	105	6·84
1-Ethyl-4-methylbenzene	105	5·22
1,2,3-Trimethylbenzene	105	6·26
1,2,4-Trimethylbenzene	105	6·35
1,3,5-Trimethylbenzene	105	6·04
n-Butylbenzene	91	10·50
Isobutylbenzene	91	10·47
s-Butylbenzene	105	7·05
t-Butylbenzene	119	0·04
1,2,3,5-Tetramethylbenzene	119	1·05
1,2,4,5-Tetramethylbenzene	119	0·90

shown in Table XXXI. Since many, but not all, phenylalkyl systems show a base peak at $m/e = 91$ ($C_7H_7^+$), an ion that contains the original phenyl nucleus—(whereas the parent molecular ion is only moderately abundant), it is the base peak rather than the molecular ion which is examined. The exceptions to this practice are benzene, which is too small to provide the $C_7H_7^+$, and certain other substituted phenyls, e.g. s-butylbenzene, which have a base peak of greater mass. Since the base peak still contains the original phenyl group, the argument is merely modified. It will be seen from Table XXXI that all the base peak ions obtained show a further loss of twenty-six units, the mass of the acetylene molecule. Such a loss varies, depending upon the nature of the alkyl group, and may be used to deduce the character of such substitutions. The abundance of this ion (base peak less 26) does not vary uniformly along the homologous series, a consideration which may be used to assign a structure to the alkyl moiety.

A tentative explanation may be advanced for this. Comparison of the homologues toluene, ethylbenzene and the isomeric butylbenzenes shows that the ion corresponding to the elimination of acetylene from the base peak falls in value along the series toluene, n-butylbenzene, isobutylbenzene ethylbenzene, s-butylbenzene, t-butylbenzene. Of these the first four cleave to produce an ion $m/e = 91$, and a great deal of evidence has been adduced to suggest that, at least in the case of toluene, it is a tropylium ion that is formed.

$C_3H_7\overset{+}{C}H_2C_6H_5 \longrightarrow$  $+ \cdot C_3H_7$

Allowing free movement of the π-bond electrons, there are seven potential acetylene residues

 $\longrightarrow C_2H_2 + C_5^+H_5$

and one may assume that these are all equally likely, so that the abundance of acetylene per carbon–carbon double bond (which may be drawn in the two canonical forms) is about 1·5% of the base peak.

For t-butylbenzene, the base peak is 119^+ ($C_9H_{11}^+$). There is no suggestion that this should re-arrange but even if it did, it could not yield a substituted tropylium ion without further modification. Accordingly one must assume that the intermediate ion (the base peak) still contains the phenyl group.

$C_6H_5^+C(CH_3)_3 \longrightarrow$ 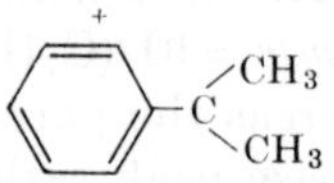 $+ CH_3$

This being so, the number of unsubstituted acetylenes which might be obtained would be at most four. Moreover, this takes no account of the possible effect of electron accession from the C_3H_6 group attached. The introduction of electrons from this group to the phenyl would lower the possibility of cleavage within the ring, and elimination of acetylene would be less probable, which accords well with the observed spectrum.

There remains the intermediate group, of which s-butylbenzene is most readily explained. The base peak is $m/e = 105$ ($C_8H_9^+$) corresponding to the fission.

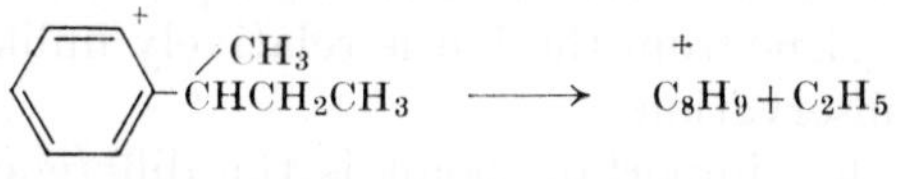

The structure of the fragment ion is uncertain, but it is almost certainly a methyl tropylium ion

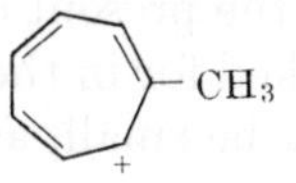

in which case there are, allowing for both canonical structures, five ways that acetylene can be eliminated, that is, about 1·5% of the base peak per double bond.

The exceptional example appears to be ethylbenzene. Here, by all the arguments already advanced, the most abundant ion $m/e = 91$ should lead to the tropylium ion, which would bring it into the same group as toluene. By similar argument, this supposes the presence of five possible ways in which acetylene could be eliminated, which would indicate a value of 1.22% of the base peak per acetylene unit, much lower than for toluene but about the same as for the trimethylbenzenes.

A study of the polymethylated benzenes shows some apparent irregularities in the ease of elimination of the C_2H_2 with structure. These variations can, however, be analysed in the same way as above. Placed in descending order of ease of elimination from the base peak, the ions group into three, the isomeric dimethylbenzenes, the isomeric trimethylbenzenes, and the tetramethylbenzenes.

Again, the preferred formulation of the bas epeak for the dimethylbenzenes $m/e = 91$ is the tropylium ion. Since this is formed in the same way for each isomer, the effect for each acetylene unit should be nearly the same for all compounds. This conclusion requires that the elimination per unit is about 0·91% of the base peak. If the structure was a substituted benzene, the value would rise to about 1.5% of the base peak per unit, which better accords with other benzene homologues. For the isomeric trimethylbenzenes, the favoured structure of the base peak ($m/e = 105$) is

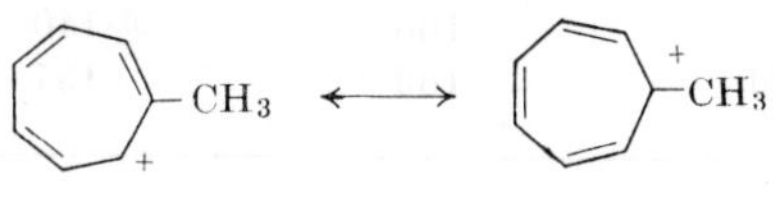

Allowing for all configurations obtainable from the tropylium structures, there are five such groups, which represent about 1·2% of the base peak per unit. With the tetramethylbenzenes, it is not possible to construct an ion for the base peak ($m/e = 119$) as a substituted tropylium ion without some consideration of which methyl is removed. The elimination of the acetylene from the ion is relatively unlikely, which is in agreement with observation.

One remaining, less important point is the difference in the ease of elimination of acetylene between the singly and multiply substituted benzenes. This is not an essential in the recognition of substituents which is the main purpose of the present discussion.

A similar effect may be looked for in the phenols and the aryl halides. In the former, the effect may be small, as the elimination of acetylene

TABLE XXXII

Compound	Base peak m/e	Base peak − 26 m/e	Base peak − 26 % Ab.
Phenol	94	68	2·32
Catechol	110	84	0·09
Resorcinol	110	84	0·15
Hydroquinone	110	84	0·44
m-Cresol	108	82	0·54
p-Cresol	107	81	2·64
2,3-Dimethylphenol	107	81	1·27
2,4-Dimethylphenol	122	96	0·32
2,5-Dimethylphenol	122	96	0·33
2,6-Dimethylphenol	122	96	0·37
3,4-Dimethylphenol	107	81	1·46
3,5-Dimethylphenol	122	96	0·32
2,4-Di-t-butylphenol	191	165	0·20
2,6-Di-t-butylphenol	191	165	0·12
2,6-Di-t-butyl hydroquinone	207	181	0·09
2,5-Di-t-amyl hydroquinone	221	195	—
2,6-Di-t-butyl-4-methylphenol	205	179	0·06
2,6-Di-t-butyl-4-ethylphenol	219	193	—
m-Methoxyphenol	124	98	—
2,4,6-Trimethylphenol	121	95	0·64
2,4,6-Tri-t-butylphenol	247	221	0·04
o-t-Butylphenol	135	109	0·97
p-t-Butylphenol	135	109	0·79
6-t-Butyl-*m*-cresol	149	123	1·20
4-t-Butylcatechol	166	140	0·19
2,6-Di-isopropylphenol	163	137	0·09

will have to compete with the known tendency of phenols to split out neutral carbon monoxide. This is indeed the case. Accordingly, in this series the loss of a further twenty-six mass units from the base peak cannot be considered diagnostically useful.

In the aryl halides, on the other hand, the effect is marked. In some

TABLE XXXIII

Compound	Base peak m/e	Base peak −26 m/e	Base peak −26 % Ab.
1-Methyl-3-fluorobenzene	109	83	13·90
1-Methyl-4-fluorobenzene	109	83	13·70
1-Methyl-2-chlorobenzene (*o*-Chlorotoluene)	91	65	9·76
1-Methyl-3-chlorobenzene (*m*-Chlorotoluene)	91	65	10·53
1-Methyl-4-chlorobenzene (*p*-Chlorotoluene)	91	65	9·53
Bromobenzene	77	51	42·50
1-Methyl-2-bromobenzene (*o*-Bromotoluene)	91	65	19·60
1-Methyl-4-bromobenzene (*p*-Bromotoluene)	91	65	20·40
1-Bromonaphthalene	206	180	0·07
5-Bromoacenaphthene	153	127	3·34
2-Bromofluorene	165	139	4·09
2,7-Dibromofluorene	243	217	0·27

instances the halogen has been eliminated to yield the base peak of the spectrum, and the problem is then essentially that of a hydrocarbon residue. In other examples, notably the two fluorotoluenes, the loss of an acetylene unit is clearly shown to persist in the presence of a halogen. The method may therefore be of diagnostic value in certain circumstances.

Appendix I

The procedure outlined in the beginning of Chapter 8 may not always be practicable. In many instances, it will not be possible to saturate olefinic or acetylenic bonds without reducing the functional groups at the same time. Here the method of analysis will have to be modified. As a general guide, the following system is suggested, but many variations may be made in the general outline.

(i) The original compound should be examined mass spectrometrically.
(ii) The functional group only should be reduced and re-examined.
(iii) The carbon skeleton should now be hydrogenated so that all functions other than aryl groups are reduced. The mass spectrum should be taken again.
(iv) The spectrum of this compound should be analysed in the usual way with further treatment with nitrous acid if needed.
(v) Comparison between the saturated system and that containing the olefinic or acetylenic groups is now made to position the centres of unsaturation.
(vi) Finally, the original structure should be deduced from the original mass spectrum.

HYDROGENATION

The wide diversity of reducing agents is such that a detailed discussion of them is beyond the scope of the present work. A few general rules may, however, be given.

(a) The reduction of double bonds in the aliphatic-carbon skeleton may be effected by hydrogen in the presence of certain finely divided metals, e.g. platinum, palladium and nickel and also by copper–chromium oxide. In many cases high temperatures and pressures may be needed. In most systems the functional groups will be reduced at the same time.

(b) Exceptionally, it is possible to reduce the double bonds in the carbon chain without affecting the carbonyl function.

(c) Hydrogenolysis of the halogen in certain alkyl halides may be achieved with palladized charcoal or Adams catalyst.

(d) Sometimes it is desirable to reduce the functional group without altering the carbon skeleton. This type of reaction may be carried out with lithium aluminium hydride in ether or some other inert solvent. This reagent will also reduce acyl halides to alcohols, an alternative to the Rosenmund reduction.

(e) In the case of polyfunctional molecules, selective reduction may be possible. Sodium borohydride will reduce carbonyl groups, but not acids.

DIAZOTIZATION

The other main reaction which has been employed in the general analytical procedure is treatment of the original or hydrogenated compound with nitrous acid. This is a rather less satisfactory reagent in that the reaction is often accompanied by re-arrangement. Thus the main product of the reaction between nitrous acid and n-butylamine is t-butyl alcohol. Similar structural alterations upon the diazotization of certain cyclic molecules occur in the Demjanov re-arrangement. Moreover, in aromatic compounds such as tribromoaniline or 2,4-dinitroaniline which possess electron withdrawing groups, the basicity of the amine is so reduced that diazotization becomes very difficult if not impossible. Further, in many such reactions it is important to avoid an excess of nitrous acid. The standard practice of removing this excess with sulphamic acid is not always possible since it may react with the compound diazotized. Also it is undesirable upon general grounds to make a complex reaction mixture, since for mass spectrometric analysis it is necessary to isolate the diazotized compound in the highest possible degree of purity.

Nevertheless, in spite of all these drawbacks, diazotization has much to commend it. It is not difficult to carry out. In favourable cases, where two amine groups are present, it may be possible to act selectively upon the more basic function. This is often realized where one amino group is in an aromatic environment and the other is aliphatic. The latter, the more basic, may be diazotized leaving the former unaffected. In the case of primary amines, the product is a primary alcohol which may be of immense importance in positioning the original functional group. The production of nitrosamines in the diazotization of secondary amines is also useful, since the addition of the NO group should be easily observed.

DESULPHURIZATION

The removal of sulphur from a compound may usually be effected by treatment with active Raney nickel. Usually, sufficient hydrogen is incorporated in the nickel by its method of preparation to effect this reaction. Thiolesters will yield aldehydes; mercaptals and mercaptols yield the hydrocarbons from which they were derived

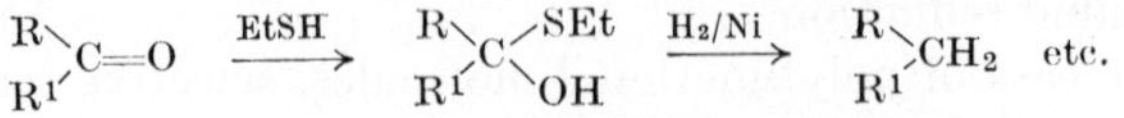

THE ROSENMUND REACTION

Hydrogenolysis of acid chlorides may be effected by hydrogenation in the presence of palladium deposited upon barium sulphate, the Rosenmund reduction. The product is the corresponding aldehyde

$$RCOCl + H_2 \rightarrow RCHO + HCl$$

The aldehyde may be further reduced to the alcohol if necessary. This reduction can be employed to produce aldehydes even when certain other reducible groups are present in the same molecule.

Appendix 2

An example of the problems found in deducing the structure of organic molecules is shown by the analysis of the mass spectra of the inositols. All eight isomers were available for the investigation. The

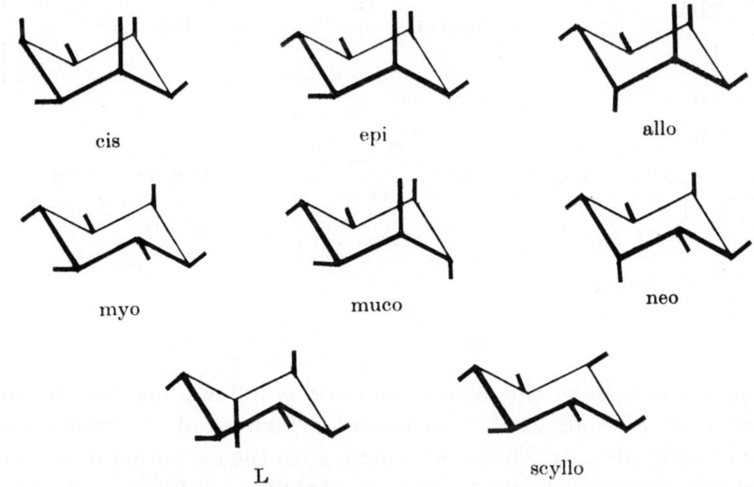

The inositols —— represent hydroxyls. The hydrogens are not shown.

inositols are relatively involatile and must be introduced upon a probe. As already explained, this technique is often associated with ion–molecule collisions, and leads to the production of a $(P+1)^+$ ion. This is important in the present circumstance, as the abundance of the parent ion is vanishingly small. Fragment ions are, however, present at $m/e = 163$, 149, 144, 102, 73, and 60. The base peaks in all spectra occur at $m/e = 73$ ($C_3H_5O_2^+$). The fragment ions $m/e = 163$ and 144 result from the elimination of one and two molecules of water, respectively, from the $(P+1)^+$ ion. The high oxygen content of the inositols makes the assignment of formulae to the remainder fairly certain, and the appropriate formulae are $m/e = 149$ ($C_5H_9O_5^+$), 102 ($C_4H_6O_3^+$), and 60 ($C_2H_4O_2^+$).

Several mechanisms can, of course, be designed. The following one is preferred for reasons to be given below.

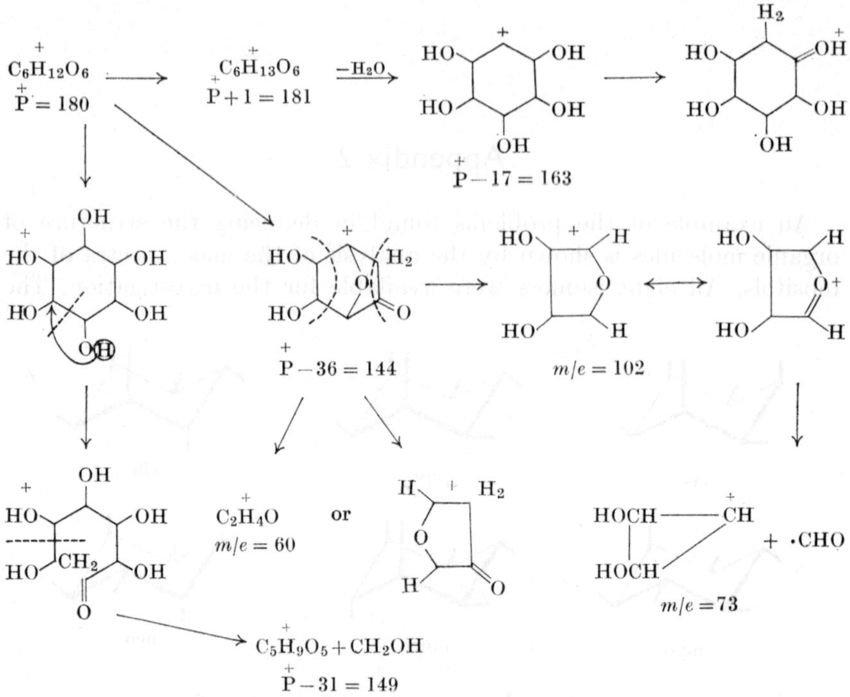

The advantages of this mechanism are as follows. (a) The cleavages invoked are all consistent with known principles of fragmentation in organic molecules. (b) The ions formed, with the exception of $m/e = 73$, are ions known to be possessed of some stability. (c) If one assumes that the stereochemical requirements for elimination in neutral molecules are retained in the ions, then the distribution of the hydroxyls will influence the ease of elimination to form given ions. (d) Employing the above argument, it is possible to predict that the elimination of two molecules of water should be more facile where the hydroxyls will be axial in the molecular ion and where the hydrogen and hydroxyl are trans-disposed to each other. By this argument, the ease of elimination may depend upon the number of favourable arrangements in the molecular ion. There are six such in cis-inositol, four in epi-, two in allo-, one in neo- and none at all in muco-inositol. This is confirmed in the mass spectra where the abundance of the appropriate ion is in the order cis > epi > neo > L > myo ~ muco.

The formation of the ions $m/e = 102$ and 73 depends upon the intermediate ion $m/e = 144$ just discussed, so that the order for these will be determined by it. This is confirmed by experiment. One further refinement is possible, namely that the ease of fission may depend upon the conformation of the hydroxyl groups near to the bridge-head carbons. This is unlikely to affect the elimination from that ion to form $m/e = 102$ since the keto-group is directed away from the hydroxyl groups on the far side of the ether linkage.

The formation of the ion $m/e = 60$ may depend upon the crowding referred to above. The formation of the *trans*-annular ether bridge gives a rigid conformation to the molecular ion. If in so doing the adjacent hydroxyl groups are brought into an axial position, the crowding of the molecular ion will be much greater than if they remain equatorial. If this consideration is valid, then the ease of elimination of $C_2H_4O_2$ should be in the order epi, neo, L, myo, allo, scyllo > muco > cis.

Upon this order is overlaid the series which determines the abundance of the intermediate ion $m/e = 144$ already given. The combination of these two factors will produce a series intermediate between those discussed.

The observed sequence is epi, allo, L, myo, neo, scyllo > muco > cis.

Appendix 3

Some of the information found in the mass spectra of the simpler organic compounds is presented tabularly in two forms. The first lists the metastable ions that have been reported in these spectra, the importance of which have already been mentioned (Chapter 2, p. 30). The metastable gives the mass of the fragment elided, while precise mass measurement of the parent and daughter ions will reveal its composition.

The second is concerned with the total ion current obtained in the electron-impact process. There is an unjustly neglected observation which relates this quantity to the ion cross-section, to which it is nearly proportional (Ötvos and Stevenson, 1955). Moreover, this current is a constitutive property of the molecule, depending upon the composition but not the structure. Strictly, the ion cross-section should take account of the cross-sections of the atoms present; however, within a homologous series, and particularly among the higher members, the observed progression does provide a rough guide which can be useful if there is no parent molecular ion. Two other relationships are included which, despite irregularities, are often useful in estimating the molecular weight and may throw some light on the molecular complexity. More exact calculations could be made, but at the present state of the subject they do not seem justified.

TRANSITIONS

15·3	$55^+ \to 29^+ + 26$		
24·0	$54^+ \to 36^+ + 18$	21·7	$85^+ \to 43 + 42$?
24·0	$96^+ \to 48^+ + 48$		
24·1	$28^+ \to 26^+ + 2$		
24·3	$30^+ \to 27^+ + 3$		
24·4	$69^+ \to 41^+ + 28$		
25·1	$29^+ \to 27^+ + 2$		
25·2	$70^+ \to 42^+ + 28$		
26·1	$30^+ \to 28^+ + 2$		
27·7	$182^+ \to 71^+ + 111$		
28·2	$54^+ \to 39^+ + 15$		
29·5	$57^+ \to 41^+ + 16$		
30·0	$56^+ \to 41^+ + 15$		
31·2	$97^+ \to 55^+ + 42$		
		31·9	$58^+ \to 43^+ + 15$
32·0	$98^+ \to 56^+ + 42$		
32·8	$59^+ \to 44^+ + 15$	33·8	$77^+ \to 51^+ + 26$
35·1	$39^+ \to 37^+ + 2$		
36·4	$83^+ \to 55^+ + 28$		

TRANSITIONS—*continued*

37·1	$41^+ \rightarrow 39^+ + 2$		
37·3	$84^+ \rightarrow 56^+ + 28$		
38·2	$44^+ \rightarrow 41^+ + 3$		
39·2	$43^+ \rightarrow 41^+ + 2$		
43·2	$70^+ \rightarrow 55^+ + 15$	45·8	$71^+ \rightarrow 57^+ + 14$
43·8	$112^+ \rightarrow 70^+ + 42$		
		46·4	$91^+ \rightarrow 65^+ + 26$
		49·0	$100^+ \rightarrow 70^+ + 30$
49·1	$97^+ \rightarrow 69^+ + 28$		
49·2	$140^+ \rightarrow 83^+ + 57$		
50·0	$98^+ \rightarrow 70^+ + 28$		
50·4	$140^+ \rightarrow 84^+ + 56$	50·4	$100^+ \rightarrow 71^+ + 29$
51·1	$55^+ \rightarrow 53^+ + 2$		
52·1			
		53·1	$57^+ \rightarrow 55^+ + 2$
56·7	$84^+ \rightarrow 69^+ + 15$	58·0	$105^+ \rightarrow 78^+ + 27$
		58·6	$86^+ \rightarrow 71 + 15$
60·0	$112^+ \rightarrow 82^+ + 30$	59·4	$105^+ \rightarrow 79^+ + 26$
61·2	$110^+ \rightarrow 82^+ + 28$		
61·5	$112^+ \rightarrow 83^+ + 29$		
		61·9	$114^+ \rightarrow 84^+ + 30$
63·0	$112^+ \rightarrow 84^+ + 28$		
		63·4	$114^+ \rightarrow 85^+ + 29$
73·3	$98^+ \rightarrow 83^+ + 15$		
		72·3	$100^+ \rightarrow 85^+ + 15$
73·1	$126^+ \rightarrow 96^+ + 30$		
74·7	$126^+ \rightarrow 97^+ + 29$		
		75·0	$128^+ \rightarrow 98^+ + 30$
75·1	$79^+ \rightarrow 77^+ + 2$		
		76·0	$78^+ \rightarrow 77^+ + 1$
76·2	$126^+ \rightarrow 98^+ + 28$		
77·1	$81^+ \rightarrow 79^+ + 2$		
		81·3	$128^+ \rightarrow 102^+ + 26$
82·3	$112^+ \rightarrow 96^+ + 16$		
84	$112^+ \rightarrow 97^+ + 15$		
		86·0	$114^+ \rightarrow 99^+ + 15$
88	$140^+ \rightarrow 111^+ + 29$	?88	$92^+ \rightarrow 90^+ + 2$
89·6	$140^+ \rightarrow 112^+ + 28$		
		90·0	$92^+ \rightarrow 91^+ + 1$
		94·9	$170^+ \rightarrow 127^+ + 43$
		99·8	$128^+ \rightarrow 113^+ + 15$
		101	$105^+ \rightarrow 103^+ + 2$
103·1	$154^+ \rightarrow 126^+ + 28$	103·4	$156^+ \rightarrow 127^+ + 29$
		105·7	$134^+ \rightarrow 119^+ + 15$
		110	$114^+ \rightarrow 112^+ + 2$
		116·9	$170^+ \rightarrow 141^+ + 29$
		125	$127^+ \rightarrow 126^+ + 1$

ALKANES

Ethane			24·1		25·1		26·1			
Propane			24·1		25·1					
n-Butane			24·1		25·1				30·4	
Isobutane										
n-Pentane				24·5	25·1			29·5		
Isopentane					25·1			29·5		
Neopentane					25·1			29·5		
n-Hexane	15·3				25·1	26·0		29·5		
2-Me-pentane	15·3				25·1	26·0		29·5		
3-Me-pentane	15·3				25·1	26·0		29·5		
2,2-DiMe-butane	15·3				25·1	26·0		29·5		
2,3-DiMe-butane	15·3				25·1	26·0				
n-Heptane	15·3	21·7			25·1	26·0		29·5		31·4
2-Me-hexane	15·3	21·7			25·1			29·5		
3-Me-hexane	15·3	21·7			25·1	26·0		29·5		
3-Et-pentane	15·3				25·1	26·0				
2,2-DiMe-pentane	15·3	21·7			25·1			29·5		
2,3-DiMe-pentane	15·3	21·7			25·1	26·0		29·5		31·4
2,4-DiMe-pentane		21·7			25·1			29·5		31·4
3,3-DiMe-pentane	15·3	21·7			25·1	26·0				
2,2,3-TriMe-butane	15·3	21·7			25·1			29·5		
n-Octane	15·3	21·7			25·1	26·0		29·5		
2-Me-heptane	15·3				25·1	26·0		29·5		
3-Me-heptane	15·3	21·7			25·1			29·5		
4-Me-heptane	15·3	21·7			25·1	26·0				
3-Et-hexane	15·3	21·7			25·1	26·0				
2,2-DiMe-hexane	15·3				25·1			29·5		
2,3-DiMe-hexane	15·3				25·1	26·0		29·5		
2,4-DiMe-hexane	15·3	21·7			25·1	26·0		29·5		
2,5-DiMe-hexane	15·3				25·1	26·0		29·5		
3,3-DiMe-hexane	15·3	21·7			25·1	26·0		29·5		
3,4-DiMe-hexane	15·3	21·7			25·1			29·5		
2-Me-3-Et-pentane	15·3	21·7			25·1	26·0		29·5		
3-Me-3-Et-pentane	15·3	21·7			25·1			29·5		
2,2,3-TriMe-pentane	15·3	21·7			25·1			29·5		
2,2,4-TriMe-pentane	15·3				25·1			29·5		
2,3,3-TriMe-pentane	15·3	21·7			25·1	26·0		29·5		
2,3,4-TriMe-pentane	15·3				25·1	26·0		29·5		
2,2,3,3-TetraMe-butane	15·3				25·1			29·5		
3-Et-heptane						26·0		29·5		
4-Et-heptane						26·0		29·5		
2,3-DiMe-heptane						26·0		29·5		
2,4-DiMe-heptane						26·0		29·5		
2,5-DiMe-heptane						26·0		29·5		
2,6-DiMe-heptane						26·0		29·5		
3,3-DiMe-heptane						26·0		29·5		

ALKANES—*continued*

3,4-DiMe-heptane				26·0	29·5
3,5-DiMe-heptane				26·0	29·5
4,4-DiMe-heptane				26·0	29·5
2-Me-3-Et-hexane				26·0	29·5
2-Me-4-Et-hexane				26·0	29·5
3-Me-3-Et-hexane				26·0	29·5
3-Me-4-Et-hexane			25·1	26·0	29·5
2,3,4-TriMe-hexane				26·0	29·5
2,3-DiMe-3-Et-pentane				26·0	29·5
2-Me-nonane	21·7		25·1	26·0	29·5
2,3-DiMe-octane		24·4	25·1	26·0	29·5
n-Undecane	21·7		25·1	26·0	29·5
3,3,5-TriMe-octane			25·1	26·0	29·5
n-Dodecane	21·7		25·1	26·0	29·5
2,2,4,6,6-PentaMe-heptane				26·0	29·5
n-Tridecane				26·0	29·5

Ethane								
Propane							39·2	
n-Butane	31·9		35·1		37·1		39·2	
Isobutane			35·1		37·1		39·2	
n-Pentane					37·1		39·2	
Isopentane					37·1			
Neopentane					37·1		39·2	
n-Hexane				36·5	37·1		39·2	
2-Me-pentane					37·1		39·2	
3-Me-pentane					37·1		39·2	
2,2-DiMe-butane					37·1		39·2	
2,3-DiMe-butane					37·1		39·2	
n-Heptane					37·1		39·2	49·0
2-Me-hexane					37·1	38·2	39·2	49·0
3-Me-hexane					37·1		39·2	49·0
3-Et-pentane					37·1		39·2	
2,2-DiMe-pentane					37·1	38·2	39·2	49·0
2,3-DiMe-pentane					37·1		39·2	
2,4-DiMe-pentane					37·1	38·2	39·2	
3,3-DiMe-pentane					37·1	38·2	39·2	
2,2,3-TriMe-butane					37·1	38·2	39·2	
n-Octane		32·8			37·1	38·2	39·2	
2-Me-heptane		32·8			37·1		39·2	
3-Me-heptane		32·8			37·1	38·2	39·2	
4-Me-heptane		32·8			37·1		39·2	
3-Et-hexane					37·1	38·2	39·2	
2,2-DiMe-hexane		32·8			37·1		39·2	

ALKANES—*continued*

2,3-DiMe-hexane		32·8	37·1			39·2					
2,4-DiMe-hexane		32·8	37·1	38·2		39·2					
2,5-DiMe-hexane		32·8	37·1			39·2					
3,3-DiMe-hexane		32·8	37·1	38·2		39·2					
3,4-DiMe-hexane		32·8	37·1	38·2		39·2					
2-Me-3-Et-pentane		32·8	37·1	38·2		39·2					
3-Me-3-Et-pentane		32·8	37·1	38·2		39·2					
2,2,3-TriMe-pentane		32·8	37·1			39·2					
2,2,4-TriMe-pentane		32·8	37·1								
2,3,3-TriMe-pentane		32·8	37·1	38·2		39·2					
2,3,4-TriMe-pentane		32·8	37·1			39·2					
2,2,3,3-TetraMe-butane		32·8	37·1			39·2					
3-Et-heptane		32·8	37·1			39·2					
4-Et-heptane		32·8	37·1	38·2		39·2					49·
2,3-DiMe-heptane			37·1	38·2		39·2					
2,4-DiMe-heptane			37·1	38·2		39·2					
2,5-DiMe-heptane		32·8	37·1			39·2					
2,6-DiMe-heptane			37·1			39·2			45·8		
3,3-DiMe-heptane		32·8	37·1			39·2			45·8		
3,4-DiMe-heptane		32·8	37·1								
3,5-DiMe-heptane		32·8	37·1								
4,4-DiMe-heptane			37·1	38·2		39·2			45·8		
2-Me-3-Et-hexane		32·8	37·1	38·2		39·2					
2-Me-4-Et-hexane		32·8	37·1			39·2					
3-Me-3-Et-hexane		32·8	37·1	38·2		39·2					
3-Me-4-Et-hexane		32·8	37·1			39·2	43·2				
2,3,4-TriMe-hexane		32·8	37·1	38·2		39·2					
2,3-DiMe-3-Et-pentane		32·8	37·1	38·2		39·2					
2-Me-nonane		32·8	37·1		39·1					48·6	
2,3-DiMe-octane	32·0	32·8	37·1		39·1					48·6	
n-Undecane		32·8	37·1	38·2		39·2		44·6			
3,3,5-TriMe-octane			37·1		39·1			44·6			
n-Dodecane		32·8	37·1	38·2		39·2			45·8		49·
2,2,4,6,6-PentaMe-heptane		32·8	37·1			39·2					
n-Tridecane					39·1			44·6			

Ethane
Propane
n-Butane
Isobutane
n-Pentane
Isopentane

ALKANES—*continued*

Neopentane													
-Hexane													
-Me-pentane													
-Me-pentane													
,2-DiMe-butane													
,3-DiMe-butane					58·6								
-Heptane		50·4											
-Me-hexane								72·3					
-Me-hexane													
-Et-pentane													
,2-DiMe-pentane													
,3-DiMe-pentane													
,4-DiMe-pentane													
,3-DiMe-pentane													
,2,3-TriMe-butane													
-Octane						61·9	63·4						
-Me-heptane											86·0		
-Me-heptane						61·9							
-Me-heptane													
-Et-hexane						61·9							
,2-DiMe-hexane													
,3-DiMe-hexane													
,4-DiMe-hexane													
,5-DiMe-hexane													
,3-DiMe-hexane											86·0		110·0
,4-DiMe-hexane						61·9	63·4						
-Me-3-Et-pentane						61·9							
-Me-3-Et-pentane													
,2,3-TriMe-pentane													
,2,4-TriMe-pentane													
,3,3-TriMe-pentane													
,3,4-TriMe-pentane													
,2,3,3-TetraMe-butane													
-Et-heptane	50·0									75·9			
-Et-heptane									75·0				
,3-DiMe-heptane													
,4-DiMe-heptane													
,5-DiMe-heptane			50·9										
,6-DiMe-heptane												99·8	
,3-DiMe-heptane				53·1									
,4-DiMe, heptane													
,5-DiMe-heptane													
,4-DiMe-heptane													
-Me-3-Et-hexane													
-Me-4-Et-hexane	50·0								75·0				
-Me-3-Et-hexane	50·0												

ALKANES—*continued*

3-Me-4-Et-hexane				75·0							
2,3,4-TriMe-hexane											
2,3-DiMe-3-Et-pentane											
2-Me-nonane	50·0										
2,3-DiMe-octane	50·0										
n-Undecane								103·4			
3,3,5-TriMe-octane											
n-Dodecane					75·1	94·9			116·9		
2,2,4,6,6-PentaMe-heptane											
n-Tridecane											

POLYCYCLIC HYDROCARBONS

Naphthalene			40·6		81·3		125·0				
Azulene			40·6		81·3		125·0				

CYCLOALKANES

Cyclopropanes

Cyclopropane	16·9	24·1					35·1				
Ethyl-											
cis-1,2-DiMe-											
trans-1,2-DiMe-											
Isopropyl-			24·4	25·1		30·0	35·1		37·3	39·1	
Et-Me-			24·4	25·1		30·0					
1,1,2-TriMe-			24·4					37·1			
1,1,2,2-TetraMe-											
Isopropenyl-				25·1	28·2		35·1	37·1			48·

Cyclopropane								
Ethyl-								
cis-1,2-DiMe-								
trans-1,2-DiMe-								
Isopropyl-	49·1	51·1	52·1					
Et-Me-								
1,1,2-TriMe-								
1,1,2,2-TetraMe-								
Isopropenyl-	49·1			54·7	63·1	75·1	77·1	

CYCLOALKANES—*continued*

				Cyclobutanes							
yclobutane	24·2			30·2							
thyl-		25·1	30·0		37·3	56·7					
				Cyclopentanes							
yclopentane					25·1		25·5				
ethyl-			24·4	24·7	25·1			30·0			
thyl-			24·4		25·1			30·0			
,1-DiMe-	15·3		24·4		25·1			30·0		32·0	35·1
is-1,2-DiMe-	15·3		24·4		25·1			30·0		32·0	35·1
ans-1,2-DiMe-	15·3		24·4		25·1			30·0		32·0	35·1
-*cis*-3-DiMe-	15·3		24·4		25·1			30·0		32·0	
-*trans*-3-DiMe-	15·3		24·4		25·1			30·0		32·0	
-Propyl-	15·3		24·4		25·1			30·0			
sopropyl-			24·4						31·2		
-Et-1-Me-	15·3		24·4		25·1				31·2		
-*cis*-Et-Me-	15·3		24·4		25·1			30·0			
,1,2-TriMe-	15·3		24·4		25·1			30·0	31·2		
,1,3-TriMe-	15·3		24·4		25·1		29·5	30·0	31·2		
-*cis*-2-*cis*-3-Tri-Me-	15·3		24·4			25·2		30·0	31·2		
-*trans*-2-*cis*-3-TriMe-	15·3		24·4			25·2		30·0	31·2		
-*cis*-2-*trans*-4-TriMe-	15·3		24·4			25·2		30·0	31·2		
-*trans*-2-*cis*-4-TriMe-	15·3		24·4			25·2		30·0	31·2		
-Butyl-			24·4		25·1			30·0		32·0	
sobutyl-			24·4		25·1		29·5	30·0			35·1
,1,3,3-TetraMe-		24·0	24·4		25·1			30·0			

yclopentane		37·1				43·2					
lethyl-			37·3	39·1							
thyl-		37·1				43·2		48·6		50·0	
,1-DiMe-	36·4	37·1				43·2		48·6			
is-1,2-DiMe-	36·4	37·1				43·2		48·6		50·0	
ans-1,2-DiMe-	36·4	37·1				43·2		48·6		50·0	
-*cis*-3-DiMe-	36·4	37·1				43·2		48·6		50·0	
-*trans*-3-DiMe-	36·4	37·1				43·2		48·6		50·0	
-Propyl-	36·4		37·3				43·8				51·1
sopropyl-		37·1			39·2		43·8		49·1		51·1
-Et-1-Me-	36·4	37·1				43·2			49·1		51·1
-*cis*-Et-Me-	36·4					43·2	43·8		49·1		51·1
,1,2-TriMe-		37·1				43·2	43·8		49·1		51·1
,1,3-TriMe-	36·4	37·1				43·2			49·1		51·1
-*cis*-2-*cis*-3-Tri-Me-	36·4	37·1				43·2	43·8		49·1		51·1
-*trans*-2-*cis*-3-TriMe-	36·4	37·1				43·2	43·8		49·1		51·1
-*cis*-2-*trans*-4-TriMe-	36·4	37·1				43·2	43·8		49·1		51·1
-*trans*-2-*cis*-4-TriMe-	36·4	37·1				43·2	43·8		49·1		51·1
-Butyl-	36·4	37·1				43·2			49·1	50·0	

CYCLOALKANES—*continued*

Isobutyl-					37·1						49·1		51
1,1,3,3-TretraMe-					37·1			43·2					51
Cyclopentane													
Methyl-			56·7										
Ethyl-						63·0		65·0	66·1				
1,1-DiMe-						63·0						75·1	
cis-1,2-DiMe-												75·1	
trans-1,2-DiMe-												75·1	
1-*cis*-3-DiMe-						63·0							
1-*trans*-3-DiMe-						63·0							
n-Propyl-				60·0	61·5	63·0			66·1			75·1	
Isopropyl-						63·0			66·1			75·1	
1-Et-1-Me-					61·5	63·0						75·1	
1-*cis*-Et-Me-				60·0	61·5	63·0						75·1	
1,1,2-TriMe-					61·5	63·0						75·1	
1,1,3-TriMe-					61·5	63·0						75·1	
1-*cis*-2-*cis*-3-Tri-Me-					61·5	63·0							
1-*trans*-2-*cis*-3-TriMe-					61·5	63·0						75·1	
1-*cis*-2-*trans*-4-TriMe-					61·5	63·0						75·1	
1-*trans*-2-*cis*-4-TriMe-					61·5	63·0						75·1	
n-Butyl-							63·1			73·1	74·7		76·
Isobutyl-							63·1					75·1	76·
1,1,3,3-TretraMe-		53·1											
Cyclohexane													
Cyclohexane	15·3	24·4		25·1			30·0					37·3	39·
Methyl-	15·3	24·4			25·2		30·0		32·0	36·4			
Ethyl-	15·3	24·4		25·1			30·0			36·4			
1,1-DiMe-	15·3	24·4		25·1			30·0	31·2			37·1		
cis-1,2-DiMe-	15·3	24·4			25·2		30·0			36·4	37·1		
trans-1,2-DiMe-	15·3	24·4		25·1			30·0	31·2		36·4	37·1		
cis-1,3-DiMe-	15·3	24·4		25·1			30·0	31·2		36·4	37·1		
trans-1,3-DiMe-	15·3	24·4		25·1			30·0	31·2		36·4	37·1		
cis-1,4-DiMe-	15·3	24·4		25·1			30·0	31·2		36·4	37·1		
trans-1,4-DiMe-	15·3	24·4		25·1			30·0	31·2		36·4	37·1		
n-Propyl-		24·4		25·1						36·4	37·1		
Isopropyl-		24·4		25·1						36·4	37·1		
1,1,3-TriMe-		24·4		25·1			30·0			36·4	37·1		
n-Butyl-		24·4		25·1						36·4	37·1		
Isobutyl-		24·4		25·1			30·0	31·2		36·4	37·1		
s-Butyl-		24·4		25·1		29·5				36·4	37·1		
t-Butyl-		24·4		25·1		29·5				36·4	37·1		
Cyclohexane	40·4											56·7	
Methyl-		43·2					50·0	51·1					
Ethyl-								51·1				60·0	61·

CYCLOALKANES—*continued*

,1-DiMe-				49·1			51·1					61·5
is-1,2-DiMe-		43·2		49·1			51·1				60·0	61·5
rans-1,2-DiMe-		43·2		49·1			51·1				60·0	61·5
is-1,3-DiMe-		43·2		49·1			51·1					
rans-1,3-DiMe-		43·2		49·1			51·1					
is-1,4-DiMe-		43·2		49·1			51·1					
rans-1,4-DiMe-		43·2		49·1			51·1					
-Propyl-				49·1			51·1	53·4	54·7			
sopropyl-							51·1	53·4	54·7			
,1,3-TriMe-		43·2					51·1					
-Butyl-			48·0		49·2		51·1					
sobutyl-			48·0		49·2		51·1					
-Butyl-			48·0		49·2		51·1					
-Butyl-							51·1					
yclohexane					75·1							
lethyl-				70·3	75·1							
thyl-	63·0				75·1							
,1-DiMe-	63·0				75·1							
is-1,2-DiMe-	63·0				75·1		84·0					
rans-1,2-DiMe-	63·0				75·1		84·0					
is-1,3-DiMe-					75·1		84·0					
rans-1,3-DiMe-	63·0				75·1	82·3	84·0					
is-1,4-DiMe-	63·0				75·1	82·3	84·0					
rans-1,4-DiMe-	63·0				75·1		84·0					
-Propyl-		63·1			75·1							
sopropyl-		63·1			75·1							
,1,3-TriMe-		63·1			75·1							
-Butyl-		63·1			75·1							
sobutyl-		63·1	67·2		75·1							
-Butyl-		63·1			75·1			86·4	88·0			
-Butyl-		63·1			75·1							

ALKENES

thene			24·1									
ropene										35·1		37·1
-Butene	15·3		24·1		25·1					35·1		37·1
is-2-Butene	15·3		24·1		25·1					35·1		37·1
rans-2-Butene	15·3		24·1		25·1					35·1		37·1
-Me-propene	15·3		24·1		25·1					35·1		37·1
-Pentene	15·3				25·1	27·7				35·1		37·1
is-2-Pentene	15·3	24·0			25·1	27·7				35·1		37·1
rans-2-Pentene	15·3	24·0			25·1	27·7				35·1		37·1
-Me-1-butene	15·3	24·0			25·1	27·7				35·1		37·1

ALKENES—*continued*

3-Me-1-butene	15·3	24·0		25·1	27·7						35·1		37·
2-Me-2-butene	15·3	24·0		25·1	27·7						35·1		37·
2,3-DiMe-1-butene			24·4										37·
2,3-DiMe-2-butene			24·4										37·
1-Hexene	15·3		24·4	25·1			30·0				35·1		37·
cis-2-Hexene	15·3		24·4	25·1	27·7		30·0				35·1		
trans-2-Hexene	15·3		24·4	25·1	27·7		30·0				35·1		37·
cis-3-Hexene	15·3		24·4	25·1	27·7		30·0				35·1		
trans-3-Hexene	15·3		24·4	25·1	27·7		30·0				35·1		37·
2-Me-1-pentene	15·3		24·4		27·7		30·0				35·1		37·
3-Me-1-pentene	15·3		24·4	25·1	27·7		30·0				35·1		37·
4-Me-1-pentene			24·4				30·0				35·1		37·
2-Me-2-pentene			24·4				30·0				35·1		37·
3-Me-*cis*-2-pentene	15·3		24·4				30·0				35·1		37·
3-Me-*trans*-2-pentene	15·3		24·4	25·1			30·0				35·1		37·
4-Me-*cis*-2-pentene			24·4				30·0						37·
4-Me-*trans*-2-pentene		24·0											37·
2-Et-1-butene	15·3		24·4	25·1	27·7		30·0				35·1		37·
3,3-DiMe-1-butene			24·4			29·5					35·1		37·
1-Heptene	15·3		24·4	25·1		29·5			32·0				
4,4-DiMe-1-pentene				25·1		29·5						36·4	37·
2,3,3-TriMe-1-butene	15·3			25·1		29·5						?36·4	37·
4,4-DiMe-*trans*-2-pentene	15·3			25·1								?36·4	
4,4-DiMe-*cis*-2-pentene	15·3		24·4	25·1								?36·4	
4-Me-1-hexene				25·1		29·5							37·
1-Octene			24·4	25·1			30·0					36·4	37·
trans-4-Octene			24·4	25·1								36·4	37·
2,4,4-TriMe-1-pentene			24·4	25·1		29·5		31·2					37·
2,4,4-TriMe-2-pentene	15·3		24·4	25·1				31·2		33·5			37·
3,4,4-TriMe-2-pentene			24·4	25·1				31·2		33·5			37·
2,3,4-TriMe-2-pentene	15·3		24·4	25·1				31·2		33·5			37·
1-Nonene			24·4	25·1		29·5		31·2				36·4	

ALKENES—*continued*

4-Nonene		24·4	25·1	26·4	28·2		30·0	31·2			36·4	
2-Me-1-octene		24·4	25·1	26·4			30·0				36·4	
4,4-DiMe-3-Et-2-pentene		24·4	25·1			29·5		31·2				37·1
1-Decene		24·4	25·1			29·5		31·2	32·0		36·4	
2-Me-3-nonene		24·4	25·1				30·0	31·2			36·4	37·1
4-Et-2-octene			25·1								36·4	
4-Et-3-octene		24·4	25·1					31·2			36·4	37·1
1-Undecene		24·4	25·1	26·4		29·5					36·4	37·1
1-Dodecene			25·1	26·4		29·5						37·1
Ethene												
Propene	38·0		40									
1-Butene												
cis-2-Butene												
trans-2-Butene												
2-Me-propene												
1-Pentene				43·2				51·1				
cis-2-Pentene				43·2				51·1				
trans-2-Pentene				43·2				51·1				
2-Me-1-butene				43·2				51·1				
3-Me-1-butene				43·2				51·1				
2-Me-2-butene				43·2				51·1				
2,3-DiMe-1-butene										56·7		
2,3-DiMe-2-butene										56·7		
1-Hexene		39·2						51·1				
cis-2-Hexene								51·1				
trans-2-Hexene								51·1			63·0	
cis-3-Hexene								51·1				
trans-3-Hexene											63·0	
2-Me-1-pentene								51·1	53·1	56·7		
3-Me-1-pentene		39·2						51·1		56·7	63·0	
4-Me-1-pentene		39·2									63·0	
2-Me-2-pentene										56·7	63·0	
3-Me-*cis*-2-pentene								51·1		56·7	63·0	
3-Me-*trans*-2-pentene						48·1		51·1		56·7		
4-Me-*cis*-2-pentene										56·7		
4-Me-*trans*-2-pentene										56·7		
2-Et-1-butene										56·7	63·0	
3,3-DiMe-1-butene		39·2			43·8	48·1				56·7	63·0	
1-Heptene				43·2			50·0				63·0	
4,4-DiMe-1-pentene											63·0	

ALKENES—*continued*

2,3,3-TriMe-1-butene									51·1		
4,4-DiMe-*trans*-2-pentene					49·1				51·1		60·0
4,4-DiMe-*cis*-2-pentene									51·1		
4-Me-1-hexene			39·2								
1-Octene				43·2							
trans-4-Octene					49·1						
2,4,4-TriMe-1-pentene					49·1				51·1		
2,4,4-TriMe-2-pentene					49·1						
3,4,4-TriMe-2-pentene					49·1						
2,3,4-TriMe-2-pentene	37·3		39·2		49·1		50·0		51·1	56·7	
1-Nonene	37·3			43·2	49·1						
4-Nonene	37·3	39·1		43·2					51·1		
2-Me-1-octene				43·2	49·1						
4,4-DiMe-3-Et-2-pentene	37·3		39·2			49·2		50·4			
1-Decene					49·1				51·1		
2-Me-3-nonene				43·2				50·4			
4-Et-2-octene		39·1		43·2				50·4			
4-Et-3-octene											
1-Undecene			39·2		49·1		50·0				
1-Dodecene		39·1					50·0				

Ethene
Propene
1-Butene
cis-2-Butene
trans-2-Butene
2-Me-propene
1-Pentene
cis-2-Pentene
trans-2-Pentene
2-Me-1-butene
3-Me-1-butene
2-Me-2-butene
2,3-DiMe-1-butene
2,3-DiMe-2-butene

ALKENES—*continued*

1-Hexene												
cis-2-Hexene												
trans-2-Hexene												
cis-3-Hexene												
trans-3-Hexene												
2-Me-1-pentene												
3-Me-1-pentene					65·0							
4-Me-1-pentene												
2-Me-2-pentene					65·0							
3-Me-*cis*-2-pentene					65·0							
3-Me-*trans*-2-pentene												
4-Me-*cis*-2-pentene												
4-Me-*trans*-2-pentene												
2-Et-1-butene				63·1	65·0							
3,3-DiMe-1-butene					65·0							
1-Heptene												
4,4-DiMe-1-pentene												
2,3,3-TriMe-1-butene	53·1						70·3		75·1			
4,4-DiMe-*trans*-2-pentene				63·1					75·1			
4,4-DiMe-*cis*-2-pentene				63·1								
2-Me-1-hexene												
1-Octene		61·5	63·0									
trans-4-Octene			63·0									
2,4,4-TriMe-1-pentene												
2,4,4-TriMe-2-pentene									75·1			
3,4,4-TriMe-2-pentene								74·7				
2,3,4-TriMe-2-pentene									75·1			
1-Nonene				63·1					75·1			
4-Nonene								74·7				
2-Me-1-octene						66·1				76·2		
4,4-DiMe-3-Et-2-pentene								74·7				
1-Decene		61·5	63·0						75·1		88·0	89·6
2-Me-3-nonene			63·0			66·1			75·1			

ALKENES—*continued*

4-Et-2-octene						66·1						
4-Et-3-octene												
1-Undecene	60·1	61·1		63·0				74·7			103·1	
1-Dodecene				63·0				74·7				

CYCLOALKENES

Cyclopentene	13·8		25·1		35·1		37·1				63·1	66·
Cyclohexene			25·1	28·4	35·1		37·1			54·7	63·1	
Methyl-cyclobutane		24·1	25·1		35·1		37·1				63·1	66·
Δ^3-*p*-Menthene						36·4	37·1	39·1	47·3			66·
Cyclopentene												
Cyclohexene		75·1	77·1									
Methyl-cyclobutane												
Δ^3-*p*-Menthene	68·3	75·1	77·1									

BICYCLIC ALKANES

cis-Decalin	24·4	25·1		37·1			49·6	54·7				
trans-Decalin	24·4	25·1		37·1		47·5	49·6	54·7			60·1	
2,3-Dihydro-indan			34·7	37·1	46·4		49·5 (49·1)		56·1	56·6		
cis-Bicyclo-[3,3,0]-octane												61·
cis-Decalin	63·1			75·1	86·1	87·7						
trans-Decalin	63·1			75·1	86·1	87·7						
2,3-Dihydro-indan		68·9	70·8				112·0	114·0	116·0			

Spiranes

Spiropentane	24·1		25·1		35·1	37·1				63·1	66·0	
Cyclopentyl-cyclopentane		24·4		28·2		37·1	46·8	48·7	60·2	63·1		
Spiropentane												
Cyclopentyl-cyclopentane	75·1		86·1	87·7								

ALKYNES

1-Butyne				24·1					28·2			
2-Butyne	13·5			24·1					28·2			
1-Pentyne		13·8				25·1						
1-Hexyne						25·1			28·2			
1-Heptyne									28·2	29·5	30·0	
1-Octyne		13·8	15·3		24·4	25·1	25·6	26·3				30·6
3-Octyne												
4-Me-1-pentyne						25·1						
3-Hexyne						25·1			28·2			
3-Me-1-butyne												
3,3-DiMe-1-butyne		13·8				25·1						
4-Octyne						25·1						
1-Nonyne						25·1						
1-Decyne						25·1				29·5		
3-Decyne						25·1						
1-Dodecyne												
2,4-Hexadiyne			19·5							32·9	33·8	34·7

1-Butyne	35·1											
2-Butyne												
1-Pentyne	35·1			37·1				49·1				
1-Hexyne	35·1			37·1	39·2			49·1				55·4
1-Heptyne	35·1			37·1	39·2			49·1				
1-Octyne	35·1	36·0		37·1	39·2		48·7		52·1			
3-Octyne				37·1	39·1							
4-Me-1-pentyne	35·1				39·2					54·7	55	
3-Hexyne	35·1					48·1				54·7		
3-Me-1-butyne				37·1				49·1				
3,3-DiMe-1-butyne	35·1			37·1								55·4
4-Octyne	34·7			37·1		47·3						
1-Nonyne			36·4	37·1	39·1							
1-Decyne		36·0	36·4	37·1	39·1							
3-Decyne				37·1	39·1	47·3						
1-Dodecyne				37·1	39·1							
2,4-Hexadiyne		39·4		48·1				59·1				

1-Butyne							
2-Butyne							
1-Pentyne			63·1	66·0			
1-Hexyne			63·1			75·1	77·1
1-Heptyne			63·1		66·1	75·1	77·1

ALKYNES—*continued*

1-Octyne				63·1		66·1						77·0			
3-Octyne	59·6			63·1	66·0					75·1				82·0	
4-Me-1-pentyne			63·0												
3-Hexyne				63·1						75·1			77·1		
3-Me-1-butyne				63·1	66·0										
3,3-DiMe-1-butyne				63·1											
4-Octyne	59·6	61·1		63·1	66·0					75·1		77·0			91·0
1-Nonyne				63·1						75·1		77·0			
1-Decyne				63·1	66·0		68·3			75·1		77·0			
3-Decyne					66·0					75·1		77·0			
1-Dodecyne					66·0		68·3			75·1		77·0			
2,4-Hexadiyne								72·1	73·0		76·0				

DIENES

1,2-Buta-	12·5	13·5	13·8		24·1						35·1	
1,3-Buta-		13·5			24·1			28·2			35·1	37·1
2-Me-1,3-buta-			13·8			25·1					35·1	37·1
3-Me-1,3-buta-			13·8			25·1					35·1	37·1
1,5-Hexa-						25·1		28·2			35·1	37·1
2,3-Penta-			13·8								35·1	37·1
1-*cis*-3-Penta-			13·8			25·1					35·1	37·1
1-*trans*-3-Penta-			13·8			25·1					35·1	37·1
1,2-Penta-			13·8			25·1					35·1	
1,4-Ponta-			13·8			25·1					35·1	
2,3-DiMe-buta-						25·1		28·2			35·1	
2,5-DiMe-1,5-hexa-				15·3		25·1	27·7					37·1
1,3-Cyclo-hexa-									28·3	34·2		

1,2-Buta-	48·1					
1,3-Buta-						
2-Me-1,3-buta-	48·1	49·1		63·1		66·0
3-Me-1,3-buta-	48·1	49·1		63·1		66·0
1,5-Hexa-	48·1		54·7	63·1		
2,3-Penta-		49·1		63·1		66·0
1-*cis*-3-Penta-		49·1		63·1		66·0
1-*trans*-3-Penta-		49·1		63·1	64·1	66·0
1,2-Penta-		49·1		63·1		66·0
1,4-Penta-		49·1				66·0

DIENES—*continued*

2,3-DiMe-buta-	—			54·7		63·1		75·1	76·1			
2,5-DiMe-1,5-hexa-	47·3	49·1	51·1		60·2		66·0	75·1			83·8	91·0
1,3-Cyclohexa-		48·1						75·1		78·0		

TERPENES

α-Pinene				37·1								
Tricyclene				37·1		46·4	49·1		64·1			
Dipentene			25·1	37·1	38·2					66·4	68·4	
Camphene				37·1			49·1		64·1	66·4		
β-Pinene		24·4		37·1		46·4			64·1			
α-Fenchene		24·4		37·1				63·6		66·4		
Cyclofenchene				37·1				63·6		66·4		
α-Pinene		71·5	75·1			89·0						
Tricyclene		71·5				89·0						
Dipentene		71·5	75·1		84·2	89·0						
Camphene		71·5	75·1		84·2	89·0						
β-Pinene		71·5	75·1		84·2	89·0						
α-Fenchene		71·5	75·1	78·0	84·2	89·0						
Cyclofenchene		71·5	75·1			89·0						

ALKYLBENZENES

Benzene								34·7	34·8			
Toluene												
Et-benzene								34·7				
n-Propylbenzene						33·8	34·0				37·1	
Isopropylbenzene							34·0					
1,2-(*o*)-Xylene							34·0	34·7		35·0		
1,3-(*m*)-Xylene							34·0	34·7		35·0		
1,4-(*p*)-Xylene							34·0	34·7		35·0		
1,2-DiMe-3-Et-			23·8		25·2		34·0			35·0	37·1	
1,2-DiMe-4-Et-			23·8		25·2		34·0			35·0	37·1	
1,3-DiMe-2-Et-			23·8		25·2		34·0				37·1	
1,3-DiMe-4-Et-			23·8		25·2		34·0				37·1	
1,3-DiMe-5-Et-			23·8		25·2		34·0				37·1	
1,4-DiMe-2-Et-			23·8		25·2		34·0			35·0	37·1	
1-Me-2-Et-						33·8						
1-Me-3-Et-						33·8					37·1	

ALKYLBENZENES—*continued*

1-Me-4-Et-				33·8					
1,2,3-TriMe-				33·8					
1,2,4-TriMe-				33·8					37·1
1,3,5-TriMe-				33·8					37·1
1,3-Di-isopropyl-					34·0				
1,4-Di-isopropyl-					34·0				
1,2-DiEt-				33·8					37·1
1,3-DiEt-		25·1		33·8					37·1
1,4-DiEt-		25·1		33·8					37·1
1-Me-3-t-Bu-									37·1
1-Me-4-t-Bu-									37·1
n-Bu-		25·1		33·8					37·1
Isobutyl-		25·1		33·8					37·1
s-Bu-		25·1		33·8					37·1
1-Me-3-isopropyl-									37·1
1-Me-4-isopropyl-									37·1
1,2,3,5-TetraMe-									37·1
1,2,4,5-TetraMe-									37·1
Styrene				33·8					

Benzene						48·1			
Toluene	43·6	46·4							
Et-benzene		46·4						57·9	58·0
n-Propylbenzene		46·4					56·5		
Isopropylbenzene		46·4	46·5						
1,2-(*o*)-Xylene		46·4						57·9	58·0
1,3-(*m*)-Xylene		46·4					56·5	57·9	58·0
1,4-(*p*)-Xylene		46·4					56·5	57·9	58·0
1,2-DiMe-3-Et-		46·4		46·8	48·0		56·5		
1,2-DiMe-4-Et-		46·4		46·8			56·5		
1,3-DiMe-2-Et-		46·4		46·8			56·5		
1,3-DiMe-4-Et-		46·4		46·8			56·5		
1,3-DiMe-5-Et-		46·4		46·8	48·0		56·5		
1,4-DiMe-2-Et-		46·4		46·8	48·0		56·5		
1-Me-2-Et-		46·4					56·5		
1-Me-3-Et-		46·4					56·5		
1-Me-4-Et-		46·4					56·5		
1,2,3-TriMe-		46·4					56·5		
1,2,4-TriMe-		46·4					56·5		
1,3,5-TriMe-		46·4					56·5		
1,3-Di-isopropyl-			46·5						
1,4-Di-isopropyl-			46·5						

ALKYLBENZENES—*continued*

1,2-DiEt-			46·4					56·5			
1,3-DiEt-			46·4					56·5			
1,4-DiEt-			46·4				49·1	56·5			
1-Me-3-t-Bu-			46·4				49·1	56·5			
1-Me-4-t-Bu-			46·4				49·1	56·5			
n-Bu-	39·1		46·4								
Isobutyl-			46·4								
s-Bu-			46·4								
1-Me-3-isopropyl-			46·4					56·5			
1-Me-4-isopropyl-			46·4					56·5			
1,2,3,5-TetraMe-			46·4				48·1	56·5			
1,2,4,5-TetraMe-			46·4					56·5			
Styrene											

Benzene									74·1		76·0
Toluene											
Et-benzene	59·4	59·5									
n-Propylbenzene	59·4				69·0					75·1	
Isopropylbenzene											
1,2-(*o*)-Xylene	59·4		59·7							75·1	
1,3-(*m*)-Xylene	59·4		59·7							75·1	
1,4-(*p*)-Xylene	59·4		59·7							75·1	
1,2-DiMe-3-Et-					69·0	69·6	70·0	71·0		75·1	
1,2-DiMe-4-Et-						69·6	70·0			75·1	
1,3-DiMe-2-Et-						69·6	70·0			75·1	
1,3-DiMe-4-Et-						69·6	70·0			75·1	
1,3-DiMe-5-Et-						69·6	70·0			75·1	
1,4-DiMe-2-Et-						69·6				75·1	
1-Me-2-Et-	59·4					69·6				75·1	
1-Me-3-Et-	59·4					69·6				75·1	
1-Me-4-Et-	59·4					69·6					
1,2,3-TriMe-	59·4					69·6				75·1	
1,2,4-TriMe-	59·4					69·6				75·1	
1,3,5-TriMe-	59·4					69·6				75·1	
1,3-Di-isopropyl-											
1,4-Di-isopropyl-											
1,2-DiEt-											
1,3-DiEt-				60·0		69·6					
1,4-DiEt-				60·0		69·6				75·1	
1-Me-3-t-Bu-				60·0		69·6				75·1	
1-Me-4-t-Bu-											
n-Bu-	59·4										
Isobutyl-						69·6					

ALKYLBENZENES—*continued*

s-Bu-		59·4	69·6		75·1
1-Me-3-isopropyl-			69·6	73·1	
1-Me-4-isopropyl-			69·6	73·1	
1,2,3,5-TetraMe-			69·6		
1,2,4,5-TetraMe-			69·6		75·1
Styrene	58·5				

Benzene												
Toluene					88·0	90·0						
Et-benzene	78·1								101·0			
n-Propylbenzene									101·0			
Isopropylbenzene								96·7				
1,2-(*o*)-Xylene	78·1				88·0				101·0			
1,3-(*m*)-Xylene	78·1				88·0				101·0			
1,4-(*p*)-Xylene	78·1				88·0				101·0			
1,2-DiMe-3-Et-				83·0							105·7	
1,2-DiMe-4-Et-		82·3		83·0							105·7	
1,3-DiMe-2-Et-		82·3		83·0							105·7	
1,3-DiMe-4-Et-											105·7	
1,3-DiMe-5-Et-		82·3		83·0							105·7	
1,4-DiMe-2-Et-		82·3		83·0							105·7	
1-Me-2-Et-							91·0		101·0			
1-Me-3-Et-							91·0		101·0			
1-Me-4-Et-							91·0		101·0			
1,2,3-TriMe-							91·0		101·0			
1,2,4-TriMe-							91·0		101·0			
1,3,5-TriMe-							91·0		101·0			
1,3-Di-isopropyl-								96·7				
1,4-Di-isopropyl-								96·7				
1,2-DiEt-		82·3										
1,3-DiEt-		82·3			88·0							
1,4-DiEt-		82·3			88·0							
1-Me-3-t-Bu-			82·9									
1-Me-4-t-Bu-			82·9									
n-Bu-					88·0	90·0						
Isobutyl-					88·0	90·0			101·0			
s-Bu-		82·3										
1-Me-3-isopropyl-												
1-Me-4-isopropyl-												
1,2,3,5-TetraMe-			82·9									
1,2,4,5-TetraMe-			82·9							105·7	113·0	114·0
Styrene												

OXYGEN COMPOUNDS

)i-isopropyl ether											
Iethanol							27·4				
:thanol											
-Propanol					25·2					29·6	
-Propanol											
-Butanol		24·2			25·2						
-Butanol					25·2				29·5		
-Me-1-propanol					25·2						
-Me-2-propanol				25·15				29·2			
-Me-1-butanol						25·5					29·7
-Butanone					25·2						
-Pentanone			25·1								
-Pentanone		24·2			25·2						
-Me-2-pentanone					25·2						29·7
-Me-1,3-penten-2-one					25·2						
-Propanol		24·2	25·1								
-Butanol			25·1								
:thyl formate			25·1								
Iethyl acetate											
:thyl acetate			25·1		25·2						
Iethyl n-propionate											
:thyl n-propionate	24·1		25·1		25·2						
Iethyl n-butyrate			25·1								
:thyl n-butyrate		24·2			25·2						

)i-isopropyl ether							37·1			39·2	
Iethanol											
:thanol									39		
-Propanol			31								
-Propanol											39·25
-Butanol		30·2								39·2	
-Butanol							37·1				
-Me-1-propanol								37·2		39·2	
-Me-2-propanol	29·8						37·1				
-Me-1-butanol										39·2	
-Butanone											
-Pentanone				31·5				37·2		39·2	
-Pentanone				31·5							
-Me-2-pentanone					32·0		37·1			39·2	
-Me-1,3-penten-2-one						36·8					
-Propanol										39·2	
-Butanol											

OXYGEN COMPOUNDS—*continued*

Ethyl-formate											
Methyl acetate											
Ethyl acetate											
Methyl n-propionate							37·1				
Ethyl n-propionate											
Methyl n-butyrate										39·2	
Ethyl n-butyrate							37·1		39·1		

Di-isopropyl ether						55					
Methanol											
Ethanol	40										
1-Propanol								58			
2-Propanol											
1-Butanol			47								
2-Butanol											
2-Me-1-propanol											
2-Me-2-propanol		43·4									
3-Me-1-butanol					54						
2-Butanone											
2-Pentanone											
3-Pentanone											
4-Me-2-pentanone											72·
4-Me-1,3-penten-2-one				51·1						70·3	
n-Propanol											
n-Butanol											
Ethyl-formate											72·0
Methyl acetate											
Ethyl acetate							56·0				
Methyl n-propionate											
Ethyl n-propionate					54						
Methyl n-butyrate											
Ethyl n-butyrate							56·0		67		

SULPHUR COMPOUNDS

(*i*)

2-Thiabutane	20·1	20·3	22·8	24·1	25·1	26·0		27·0	29·1		
3-Thiapentane	20·1	20·3	22·8	24·1	25·1						29
3-Thiahexane			22·8	24·1	25·1	26·0		27·0		29·4	29
3-Thiaheptane				24·1	25·1	26·0	26·6	27·0			29

SULPHUR COMPOUNDS—*continued*

Thiacyclobutane				24·1							
Thiacyclopentane											
-Thiabutane	30·3						35·0				
-Thiapentane						34·3					
-Thiahexane				34·0	34·1		35·0	37·1	38·1	39·1	
-Thiaheptane		32·7									
hiacyclobutane								37·1			
hiacyclopentane			33·1					37·1			41·0
-Thiabutane		45·0	46·0	47·0	48·2						
-Thiapentane	42·7										
-Thiahexane	42·7	45·0		47·0			58·0	62·0			
-Thiaheptane											
Thiacyclobutane											
Thiacyclopentane						55·1					
				(ii)							
Ethanethiol					24·1	25·1			26·1	27·0	28·0
-Pentanethiol		20·1				25·1			26·1	27·0	
-Bu-mercapton			22·4	22·8		25·1		26·0		27·0	
,3-Dithiabutane							25·6				
,4-Dithiahexane		20·1			24·1	25·1					
Ethanethiol				34·1							
-Pentanethiol		29·1	29·5		37·1			43·2		51·1	
-Bu-mercapton			29·5		37·1						
,3-Dithiabutane							39·6				
,4-Dithiahexane						37·4		46·3	48·2		72·4

THIOPHENES

hiophene				34·9	35·1		37·0	37·1			
-Me-thiophene				34·9							
-Me-thiophene		27·2		34·9							39·1
-Et-thiophene				34·9							

THIOPHENES—*continued*

2,3-DiMe-thiophene					34·9						
2,4-DiMe-thiophene											
2,5-DiMe-thiophene											
2-t-Bu-thiophene								37·1			
3-t-Bu-thiophene											
Tetrahydrothiophene						35·1	37·0			39·0	
2-Thiophenthiol	14·1		28·8	34·6				37·1			39·

Thiophene		40·0									
2-Me-thiophene									50·0		53·
3-Me-thiophene									50·0	51·0	
2-Et-thiophene											
2,3-DiMe-thiophene											
2,4-DiMe-thiophene											
2,5-DiMe-thiophene											
2-t-Bu-thiophene											
3-t-Bu-thiophene											
Tetrahydrothiophene		40·0	41·4		45·0		46·1	47·1			
2-Thiophenthiol	39·2			44·2		45·4					

Thiophene		58·0	58·1								
2-Me-thiophene						63·0					
3-Me-thiophene								67·0			
2-Et-thiophene											
2,3-DiMe-thiophene											
2,4-DiMe-thiophene											
2,5-DiMe-thiophene											
2-t-Bu-thiophene			58·1				66·8				
3-t-Bu-thiophene											
Tetrahydrothiophene	55·0										
2-Thiophenthiol				59·4	60·0				71·0		

Thiophene					
2-Me-thiophene				96·0	
3-Me-thiophene				96·0	
2-Et-thiophene			84·8		
2,3-DiMe-thiophene	75·1		84·8		
2,4-DiMe-thiophene	75·1		84·8		
2,5-DiMe-thiophene	75·1				
2-t-Bu-thiophene		75·3			111·6
3-t-Bu-thiophene					

ALKANES

	Mol. wt.	% Parent ion (P)	Abundance Σ	$\frac{100P}{\Sigma}$	$\frac{100MP}{\Sigma}$
Methane	16	100	218·52	45·76	732·0
Ethane	30	26·2	439·59	5·96	178·8
Propane	44	29·2	812·22	3·60	158·2
n-Butane	58	12·6	1124·67	1·12	65·0
n-Pentane	72	8·84	1450·55	0·61	47·6
n-Hexane	86	14·1	2005·84	0·70	60·5
n-Heptane	100	13·0	474·92	2·74	273·7
n-Octane	114	6·92	963·12	0·72	81·9
n-Nonane	128	6·38	1393·81	0·46	58·9
n-Decane	142	6·03	1855·40	0·33	46·9
n-Undecane	156	3·80	468·56	0·81	126·5
n-Dodecane	170	4·16	960·62	0·43	73·4
n-Tridecane	184	4·56	1938·35	0·24	44·2
n-Tetradecane	198	3·40	2814·29	0·12	24·0
n-Pentadecane	212	2·93	3595·73	0·08	17·2
n-Hexadecane	226	2·53	4548·68	0·056	12·6
n-Heptadecane	240	2·05	866·92	0·24	57·6
n-Octadecane	254	1·71	8196·20	0·20	50·8
n-Eicosane	282	1·38	1354·53	0·10	28·2
n-Heneicosane	296	1·61	951·61	0·17	50·32
n-Docosane	310	1·47	981·99	0·15	46·50
n-Tricosane	324	2·17	956·00	0·23	74·52
n-Tetracosane	338	7·36	600·28	1·22	412·36
n-Pentacosane	352	1·67	960·52	0·17	59·84
n-Hexacosane	366	1·20	766·98	0·16	58·56
n-Heptacosane	380	1·23	961·76	0·13	49·40
n-Octacosane	394	1·28	405·35	0·32	126·08
n-Nonacosane	408	1·01	1014·53	0·10	40·8
n-Triacontane	422	0·85	971·22	0·09	37·98
n-Dotriacontane	450	0·59	407·66	0·14	63·00
n-Hexatriacontane	506	0·46	407·97	0·11	55·66
n-Tetratetracontane	618	0·28	473·61	0·06	37·14

ALKANES—*continued*

	Mol. wt.	% Parent ion (P)	Abundance Σ	100P/Σ	100MP/Σ
2-Me-3-Et-hexane	128	0·53	342·64	0·15	19·85
2-Me-4-Et-hexane	128	0·92	382·94	0·24	30·72
3-Me-3-Et-hexane	128	0·02	371·86	0·01	0·69
3-Me-4-Et-hexane	128	1·52	559·78	0·27	34·84
3,3-DiEt-pentane	128	0	261·06	0	0
2,2,3,3-TetraMe-pentane	128	0	388·83	0	0
2,2,3,4-TetraMe-pentane	128	0·02	380·98	0	0·67
2,2,4,4-TetraMe-pentane	128	0·02	219·33	0·91	1·17
2,3,3,4-TetraMe-pentane	128	0	304·44	0	0
n-Decane	142	6·03	1855·40	0·32	4·62
2-Me-nonane	142	1·36	472·32	0·29	4·09
2,3-DiMe-octane	142	0·85	386·53	0·22	3·13
2,2,4-TriMe-heptane	142	0	2921·96	0	0
3,3,5-TriMe-heptane	142	0·03	439·85	0·01	0·10
2-Me-5-Et-heptane	142	0	5288·99	0	0
n-Butane	58	12·6	1124·67	1·12	65·0
Isobutane	58	2·75	1607·84	0·17	9·9
n-Pentane	72	8·84	1450·55	0·61	43·9
Isopentane	72	2·75	1607·84	0·17	12·24
Neopentane	72	0·01	240·81	0	0
n-Hexane	86	14·1	2005·84	0·70	60·5
Isohexane	86	2·98	332·17	0·90	77·4
3-Me-pentane	86	3·01	440·55	0·68	58·74
2,2-DiMe-butane	86	0·05	549·07	0·01	0·9
2,3-DiMe-butane	86	3·43	311·37	1·10	94·8
n-Heptane	100	13·0	474·92	2·74	273·7
Isoheptane	100	3·86	361·46	1·07	106·8
3-Me-hexane	100	3·67	477·08	0·77	77·0
3-Et-pentane	100	1·79	331·08	0·54	54·17
2,2-DiMe-pentane	100	0·05	406·87	0·01	1·0
2,3-DiMe-pentane	100	2·22	508·36	0·44	44·00
2,4-DiMe-pentane	100	1·14	392·92	0·29	29·06
3,3-DiMe-pentane	100	0·01	324·32	0	0·03
2,2,3-TriMe-butane	100	0·03	452·91	0·01	0·66
n-Octane	114	6·92	963·12	0·72	81·9
Iso-octane	114	4·91	421·86	1·16	132·7

ALKANES—*continued*

	Mol. wt.	% Parent ion (P)	Abundance Σ	100P/Σ	100MP/Σ
3-Me-heptane	114	2·95	471·92	0·63	71·3
4-Me-heptane	114	3·08	362·65	0·85	96·8
3-Et-hexane	114	1·64	309·27	0·53	60·42
2,2-DiMe-hexane	114	0·03	242·35	0·01	1·24
2,3-DiMe-hexane	114	1·67	384·79	0·43	49·56
2,4-DiMe-hexane	114	1·69	454·65	0·37	42·45
2,5-DiMe-hexane	114	3·83	416·00	0·92	105·14
3,3-DiMe-hexane	114	0·01	389·91	0	0·29
3,4-DiMe-hexane	114	2·15	466·46	0·46	52·44
2-Me-3-Et-pentane	114	1·25	347·81	0·36	41·04
3-Me-3-Et-pentane	114	0	326·54	0	0
2,2,3-TriMe-pentane	114	0·03	281·93	0·01	1·22
2,2,4-TriMe-pentane	114	0·02	249·12	0·01	0·92
2,3,3-TriMe-pentane	114	0·01	374·47	0·03	3·05
2,3,4-TriMe-pentane	114	0·33	337·46	0·10	11·40
2,2,3,3-TetraMe-butane	114	0·03	237·19	0·01	1·44
n-Nonane	128	6·38	1393·81	0·46	58·4
Isononane	128	2·26	377·68	0·60	76·54
3-Me-octane	128	1·64	318·70	0·52	65·92
4-Me-octane	128	2·41	351·91	0·69	87·7
3-Et-heptane	128	1·17	370·39	0·32	40·53
4-Et-heptane	128	1·30	365·80	0·36	45·60
2,2-DiMe-heptane	128	0·06	299·55	0·02	2·57
2,3-DiMe-heptane	128	1·03	307·78	0·33	42·24
2,4-DiMe-heptane	128	0·65	331·14	0·20	25·18
2,5-DiMe-heptane	128	1·03	338·31	0·30	39·06
2,6-DiMe-heptane	128	2·73	354·63	0·77	98·77
3,3-DiMe-heptane	128	0	537·69	0	0
3,4-DiMe-heptane	128	1·21	509·10	0·24	30·74
3,5-DiMe-heptane	128	0·63	310·84	0·20	26·00
4,4-DiMe-heptane	128	0	354·96	0	0
2,2,3-TriMe-hexane	128	0·03	312·07	0·01	1·23
2,2,4-TriMe-hexane	128	0·03	266·01	0·01	1·45
2,2,5-TriMe-hexane	128	0·06	270·92	0·02	2·84
2,3,3-TriMe-hexane	128	0	364·46	0	0
2,3,5-TriMe-hexane	128	1·49	324·44	0·46	58·92
2,4,4-TriMe-hexane	128	0	491·96	0	0
3,4,4-TriMe-hexane	128	0·03	563·07	0·01	0·68

CYCLOALKANES*

	Mol. wt.	% Parent ion (P)	Abundance Σ	$\frac{100P}{\Sigma}$	$\frac{M \times P^*}{\Sigma}$
Cyclopropane	42	100	400·42	24·97	10·51
Cyclobutane	56	62·20	418·00	14·88	8·33
Cyclopentane	70	29·5	266·93	11·05	7·75
1-*cis*-2-DiMe-cyclopropane	70	50·10	432·05	11·60	8·12
1-*trans*-2-DiMe-cyclopropane	70	53·80	416·78	12·91	8·34
Cyclohexane	84	70·5	482·43	14·61	12·27
Me-cyclopentane	84	15·4	393·66	3·91	3·28
Et-cyclobutane	84	2·71	330·38	0·82	0·69
Isopropylcyclopropane	84	1·73	453·84	0·50	0·32
1-Et-1-Me-cyclopropane	84	32·20	528·38	6·09	5·12
1,1,2-TriMe-cyclopropane	84	29·7	320·63	9·26	7·79
Cycloheptane	98	64·20	909·99	7·06	6·92
Me-cyclohexane	98	44·0	577·64	7·62	7·47
Et-cyclopentane	98	16·5	704·98	2·34	2·30
1,1-DiMe-cyclopentane	98	8·15	601·00	1·36	1·34
1-*cis*-2-DiMe-cyclopentane	98	22·4	639·29	3·50	3·44
1-*trans*-2-DiMe-cyclopentane	98	25·7	553·77	4·64	4·56
1-*cis*-3-DiMe-cyclopentane	98	16·0	634·74	2·52	2·47
1-*trans*-3-DiMe-cyclopentane	98	14·5	644·73	2·25	2·21
1,1,2,2-TetraMe-cyclopropane	98	23·10	510·30	4·53	4·44
Cyclo-octane	112	42·90	868·43	5·00	5·53
Et-cyclohexane	112	20·2	447·52	4·51	5·06
1,1-DiMe-cyclohexane	112	5·32	498·09	1·07	1·20
1-*cis*-2-DiMe-cyclohexane	112	31·70	625·79	5·07	5·69
1-*trans*-2-DiMe-cyclohexane	112	33·90	617·51	5·49	6·16
1-*cis*-DiMe-cyclohexane	112	29·50	464·25	6·35	7·11
1-*trans*-3-DiMe-cyclohexane	112	36·10	529·34	6·82	7·65
1-*cis*-4-DiMe-cyclohexane	112	30·9	495·26	6·24	7·00
1-*trans*-4-DiMe-cyclohexane	112	34·8	496·5	7·01	7·85
n-Propylcyclopentane	112	16·6	753·03	2·20	2·47
Isopropylcyclopentane	112	2·87	555·50	0·52	0·58
1-Me-1-Et-cyclopentane	112	0·97	463·49	0·21	0·24
1-Me-*cis*-2-Et-cyclopentane	112	15·1	778·8	1·94	2·18
1-Me-*trans*-3-Et-cyclopentane	112	87·10	6776·88	1·29	1·44
1,1,2-TriMe-cyclopentane	112	14·0	658·83	2·12	2·38
1,1,3-TriMe-cyclopentane	112	7·0	747·82	0·09	0·10
1-*cis*-2-*cis*-3-TriMe-cyclopentane	112	6·91	4·42	1·56	1·75

* Note change in units.

CYCLOALKANES—*continued*

	Mol. wt.	% Parent ion (P)	Abundance Σ	100P/Σ	M×P/Σ
1-*trans*-2-*cis*-3-TriMe-cyclopentane	112	18·3	636·0	2·88	3·23
1-*cis*-2-*trans*-4-TriMe-cyclopentane	112	6·20	354·59	1·75	1·96
1-*trans*-2-*cis*-4-TriMe-cyclopentane	112	9·49	397·90	2·39	2·68
n-Propylcyclohexane	126	20·3	446·91	4·54	5·73
Isopropylcyclohexane	126	15·8	545·8	2·89	3·65
1,1,3-TriMe-cyclohexane	126	23·8	480·63	4·95	6·25
1,2,3-TriMe-cyclohexane	126	3·12	8004·01	0·04	0·05
1,2,4-TriMe-cyclohexane	126	3·08	7941·00	0·04	0·05
n-Butylcyclopentane	126	17·0	789·15	2·15	2·71
Isobutylcyclopentane	126	6·56	817·89	0·80	1·01
1,1,3,3-TetraMe-cyclopentane	126	0·35	693·46	0·05	0·06
n-Butylcyclohexane	140	15·0	444·85	3·37	4·73
Isobutylcyclohexane	140	14·60	563·39	2·59	3·63
s-Butylcyclohexane	140	14·90	709·57	2·10	2·95
t-Butylcyclohexane	140	0·96	365·79	0·26	0·36
1-Me-*cis*-4-isopropyl-cyclohexane	140	8·98	457·16	1·96	2·74
1-Me-*trans*-4-isopropyl-cyclohexane	140	11·90	472·21	2·52	3·53
2-Cyclohexyl-2-Me-butane	154	0·22	464·00	0·05	0·08

MISCELLANEOUS CYCLIC ALKANES

	Mol. wt.	% Parent ion (P)	Abundance Σ	100P/Σ	M×P/Σ
Spiropentane	68	14·9	501·22	2·97	2·02
cis-Bicyclo-(3,3,0)-octane	110	14·59	378·22	3·86	4·25
cis-Hexahydro-indane	124	32·70	826·46	3·96	4·91
Tricyclo-(3,1,1,3,7)-decane	136	100·00	603·80	16·56	22·52
Tricyclene	136	21·20	435·95	4·86	6·61

MISCELLANEOUS CYCLIC ALKANES—*continued*

	Mol. wt.	% Parent ion (P)	Abundance Σ	$\frac{100P}{\Sigma}$	$\frac{M \times P}{\Sigma}$
cis-Decahydronaphthalene	138	72·60	995·62	7·29	10·06
trans-Decahydronaphthalene	138	87·40	1032·08	8·47	11·69
Cyclopentylcyclopentane	138	10·80	719·66	1·50	2·07
1,1-Dicyclopentylethane	166	1·02	466·57	0·22	0·36
1,3-Dicyclopentylcyclopentane	206	6·64	816·95	0·81	1·67
1,5-Dicyclohexyl-3-(3′-cyclopentylpropyl)pentane	346	0·10	689·80	0·01	0·35
1,7-Dicyclopentyl-4-(2′-cyclohexylethyl)heptane	346	0·40	849·23	0·05	0·17

ALKENES

	Mol. wt.	% Parent ion (P)	Abundance Σ	$\frac{100P}{\Sigma}$	$\frac{M \times P}{\Sigma}$
Ethylene	28	100	262·47	3·81	1·07
Propene	42	100	383·11	2·61	1·10
1-Butene	56	37·00	323·09	11·45	6·41
cis-2-Butene	56	47·90	360·39	13·29	7·44
trans-2-Butene	56	46·20	357·94	12·91	7·23
2-Me-propene	56	42·80	320·94	13·34	7·47
1-Pentene	70	31·70	392·19	8·08	5·66
cis-2-Pentene	70	33·60	367·84	9·13	6·39
trans-2-Pentene	70	33·40	363·01	9·20	6·44
2-Me-1-butene	70	31·00	343·55	9·02	6·31
3-Me-1-butene	70	26·20	314·86	8·32	5·82
2-Me-2-butene	70	35·70	354·66	10·06	7·04
1-Hexene	84	28·20	624·56	4·52	3·80
cis-2-Hexene	84	27·90	508·32	5·49	4·61
trans-2-Hexene	84	28·90	428·20	6·75	5·67
cis-3-Hexene	84	36·80	549·24	6·70	5·63
trans-3-Hexene	84	36·90	540·38	6·83	5·74
2-Me-1-pentene	84	31·30	406·60	7·71	6·48
3-Me-1-pentene	84	29·20	559·58	5·22	4·38
4-Me-1-pentene	84	10·70	404·75	2·64	2·22
2-Me-2-pentene	84	30·30	385·87	7·85	6·59

ALKENES—*continued*

	Mol. wt.	% Parent ion (P)	Abundance Σ	100P/Σ	M × P/Σ
3-Me-*cis*-2-pentene	84	36·00	462·79	7·78	6·54
3-Me-*trans*-2-pentene	84	37·30	467·50	7·98	6·70
4-Me-*cis*-2-pentene	84	28·70	378·43	7·58	6·37
4-Me-*trans*-2-pentene	84	27·70	370·48	7·48	6·28
2-Et-1-butene	84	40·90	522·69	7·82	6·57
2,3-DiMe-1-butene	84	23·30	348·70	6·68	5·61
3,3-DiMe-1-butene	84	20·70	364·81	5·67	4·76
2,3-DiMe-2-butene	84	29·80	375·25	7·94	6·67
1-Heptene	98	16·40	670·41	2·45	2·41
trans-2-Heptene	98	41·90	669·65	6·26	6·13
trans-3-Heptene	98	27·90	507·71	5·50	5·39
2-Me-1-hexene	98	4·39	334·74	1·31	1·28
3-Me-1-hexene	98	10·20	538·73	1·89	1·85
4-Me-1-hexene	98	4·97	537·92	0·92	0·90
5-Me-1-hexene	98	1·71	572·94	0·30	0·29
2-Me-2-hexene	98	25·00	437·52	0·57	0·56
3-Me-*cis*-2-hexene	98	31·80	507·94	6·26	6·13
4-Me-*trans*-2-hexene	98	19·60	417·62	4·69	4·60
5-Me-2-hexene	98	14·10	581·53	2·42	2·37
2-Me-*trans*-3-hexene	98	28·60	553·76	5·16	5·06
3-Me-*cis*-3-hexene	98	28·20	445·55	6·33	6·22
3-Me-*trans*-3-hexene	98	29·00	521·64	5·56	5·45
3-Et-1-pentene	98	16·50	446·13	3·70	3·63
3-Et-2-pentene	98	38·30	595·98	6·43	6·30
2,3-DiMe-1-pentene	98	14·20	509·00	2·79	2·73
2,4-DiMe-1-pentene	98	7·76	396·43	1·96	1·92
3,3-DiMe-1-pentene	98	8·96	435·75	2·06	2·01
3,4-DiMe-1-pentene	98	0·82	640·61	0·13	1·27
4,4-DiMe-1-pentene	98	1·59	320·62	0·50	4·90
2,3-DiMe-2-pentene	98	38·50	572·64	6·72	6·59
2,4-DiMe-2-pentene	98	26·40	447·52	5·90	5·78
3,4-DiMe-*cis*-2-pentene	98	34·50	504·85	6·83	6·69
3,4-DiMe-*trans*-2-pentene	98	34·50	506·36	6·81	6·67
4,4-DiMe-*cis*-2-pentene	98	26·90	504·73	5·33	5·22
4,4-DiMe-*trans*-2-pentene	98	27·20	448·91	6·06	5·94
3-Me-2-Et-1-butene	98	30·20	661·11	4·57	4·48

ALKENES—*continued*

	Mol. wt.	% Parent ion (P)	Abundance Σ	100P / Σ	M × P / Σ
2,3,3-TriMe-1-butene	98	20·30	445·12	4·51	4·47
1-Octene	112	10·80	798·56	1·35	1·51
trans-4-Octene	112	23·20	593·97	3·91	4·38
2-Me-*trans*-3-heptene	112	22·90	576·05	3·98	4·46
2,3-DiMe-2-hexene	112	29·10	549·51	5·30	5·94
2,5-DiMe-2-hexene	112	19·60	382·90	5·12	5·73
2,2-DiMe-*cis*-hexene	112	15·20	483·56	3·14	3·52
2,2-DiMe-*trans*-3-hexene	112	18·20	477·09	3·81	4·27
2,5-DiMe-*trans*-3-hexene	112	17·80	462·41	3·85	4·31
2-Me-3-Et-1-pentene	112	6·90	387·32	1·78	1·99
2,4,4,TriMe-1-pentene	112	9·73	304·18	3·20	3·58
2,3,4-TriMe-2-pentene	112	30·40	530·28	5·73	6·42
2,4,4-TriMe-2-pentene	112	25·30	500·53	5·05	5·66
3,4,4-TriMe-2-pentene	112	26·60	468·07	5·68	6·36
1-Nonene	126	7·34	729·13	1·01	1·27
4-Nonene	126	19·40	545·22	3·56	4·49
2-Me-1-octene	126	6·67	371·57	1·79	2·27
3,4,4-TriMe-2-hexene	126	15·50	465·88	3·33	3·73
4,4-DiMe-3-Et-2-pentene	126	27·10	716·27	3·77	4·22
1-Decene	140	6·42	861·90	0·75	1·04
2-Me-3-nonene	140	14·10	651·56	2·16	3·02
4-Et-2-octene	140	9·98	453·10	2·20	3·08
4-Et-3-octene	140	16·6	568·11	2·92	4·09
2,2,5,5-TetraMe-*cis*-3-hexene	140	1·34	924·00	0·15	0·21
2,2,5,5-TetraMe-*trans*-3-hexene	140	7·72	791·43	0·98	1·37
1-Undecene	154	4·00	867·37	0·46	0·71
1-Dodecene	168	2·15	838·22	0·26	0·44
1-Tridecene	182	4·72	1591·55	0·30	0·55

ALKENES—*continued*

	Mol. wt.	% Parent ion (P)	Abundance Σ	$\frac{100P}{\Sigma}$	$\frac{M \times P}{\Sigma}$
1-Tetradecene	196	4·04	1581·48	0·26	0·51
1-Pentadecene	210	3·53	1422·49	0·25	0·53
1-Hexadecene	224	3·79	1592·85	0·24	0·54

ALCOHOLS

	Mol. wt.	% Parent ion (P)	Abundance Σ	$\frac{100P}{\Sigma}$	$\frac{M \times P}{\Sigma}$
Methanol	32	100·00	232·20	43·07	13·78
Ethanol	46	7·30	253·38	2·88	13·25
1-Propanol	60	10·50	228·17	0·46	0·28
2-Propanol	60	0·51	181·85	0·28	0·17
1-Butanol	74	0·94	536·80	0·18	0·13
2-Butanol	74	0·41	245·29	0·17	0·12
2-Methyl-1-propanol	74	12·70	457·43	2·78	2·05
2-Methyl-2-propanol	74		293·41		
1-Pentanol	88		573·29		
3-Methyl-1-butanol	88		714·47		
2-Methyl-2-butanol	88		579·70		
2-Pentanol	88	0·04	203·24	0·02	0·02
3-Pentanol	88	0·06	266·54	0·02	0·02
2-Methyl-1-butanol	88	0·20	195·45	0·10	0·09
2-Methyl-2-butanol	88	0·31	584·44	0·05	0·04
1-Hexanol	102		615·36		
2-Methyl-1-pentanol	102		436·73		
3-Methyl-1-pentanol	102		793·85		
4-Methyl-2-pentanol	102	0·07	259·13	~0	0·02
2-Ethyl-1-butanol	102		444·67		
2-Heptanol	116	0·01	293·70	~0	~0
3-Heptanol	116	0·02	529·94	~0	~0
4-Heptanol	116	0·02	418·45	~0	0·01
1-Octanol	130		1008·89		
2-Octanol	130	0·01	278·89	~0	0·03
5-Et-2-heptanol	144		939·80		
9-Heptadecanol	256	0·12	794·50	0·02	0·04

ALCOHOLS—*continued*

	Mol. wt.	% Parent ion (P)	Abundance Σ	$\frac{100P}{\Sigma}$	$\frac{M \times P}{\Sigma}$
1-Decanol	158	0·02	844·97	~0	~0
1-Dodecanol	186		909·45		
1-Tetradecanol	214	0·01	915·71	~0	~0
1-Hexadecanol	242	0·01	891·66	~0	~0
1-Octadecanol	270	0·01	838·78	~0	~0

ESTERS

	Mol. wt.	% Parent ion (P)	Abundance Σ	$\frac{100P}{\Sigma}$	$\frac{M \times P}{\Sigma}$
Methyl formate	60	27·94	248·25	11·26	6·74
Ethyl formate	74	7·07	366·77	1·93	1·43
n-Propyl formate	88	0·33	306·62	0·11	0·10
n-Butyl formate	102	0·33	493·32	0·07	0·07
Isobutyl formate	102	0·46	572·41	0·07	0·07
s-Butyl formate	102	0·17	400·54	0·04	0·04
Methyl acetate	74	15·21	157·09	9·68	7·17
Ethyl acetate	88	3·95	197·21	2·00	1·76
n-Propyl acetate	102	0·04	220·77	0·02	0·02
Isopropyl acetate	102	0·11	224·57	0·05	0·05
n-Butyl acetate	116	0·02	254·35	0·01	0·01
Methyl propionate	88	21·29	326·21	6·53	5·75
Ethyl propionate	102	6·91	286·72	2·41	2·46
n-Propyl propionate	116	0·02	346·77	0·01	0·01
Isopropyl propionate	116	0·22	418·31	0·05	0·06
n-Butyl propionate	130	0·02	366·70	0·01	0·01
Isobutyl propionate	130	0·04	297·52	0·01	0·01
Methyl butyrate	102	1·86	467·11	0·40	0·41
Ethyl butyrate	116	4·36	606·70	0·72	0·84
n-Propyl butyrate	130	0·05	497·87	0·01	0·01
n-Butyl n-butyrate	144	0·07	609·42	0·01	0·01
Methyl isobutyrate	102	13·00	334·17	3·89	3·97
Isobutyl isobutyrate	144	0·03	515·63	0·01	0·01
Methyl pivalate	116	3·76	287·49	1·31	1·52

ALDEHYDES

	Mol. wt.	% Parent ion (P)	Abundance Σ	$\frac{100P}{\Sigma}$	$\frac{M \times P}{\Sigma}$
Methanol	32	88·50	235·00	37·66	12·05
Ethanol	46	45·70	212·15	21·54	9·91
Propanol	60	37·38	348·78	10·72	6·43
n-Butanol	74	44·75	631·93	7·08	5·24
2-Me-1-propanol	88	35·50	409·47	8·70	7·63
3-Me-1-butanol	88	10·0	746·28	1·35	1·17

KETONES

	Mol. wt.	% Parent ion (P)	Abundance Σ	$\frac{100P}{\Sigma}$	$\frac{M \times P}{\Sigma}$
2-Propanone	58	22·65	171·27	13·23	7·68
2-Butanone	72	17·01	193·33	8·80	6·34
2-Pentanone	86	14·95	240·70	6·21	5·35
3-Pentanone	86	17·32	312·29	5·55	4·78
3-Me-2-butanone	86	9·81	166·90	5·88	5·06
2-Hexanone	100	6·40	255·35	2·51	2·51
3-Hexanone	100	19·90	457·07	4·35	4·35
3-Me-2-pentanone	100	3·46	269·92	1·28	1·28
4-Me-2-pentanone	100	10·54	261·09	4·04	4·04
2-Heptanone	114	2·94	241·78	1·22	1·39
3-Heptanone	114	7·78	325·97	2·39	2·72
2,4-DiMe-3-pentanone	114	5·94	223·12	2·66	3·04
2,6-DiMe-4-heptanone	142	11·19	479·60	2·33	3·32
5-Nonanone	142	11·05	547·32	2·02	2·87
11-Heneicosanone	310	2·88	933·00	0·31	0·96

SULPHUR COMPOUNDS

	Mol. wt.	% Parent ion (P)	Abundance Σ	$\frac{100P}{\Sigma}$	$\frac{M \times P}{\Sigma}$
Thiacyclobutane	74	43·20	225·33	19·17	14·19
Thiacyclopentane	88	53·10	452·13	11·74	10·33
Thiacyclopentane	88	53·30	436·47	12·21	10·74
3-Me-thiacyclobutane	88	41·50	370·67	11·20	9·86
Thiacyclohexane	102	100·00	2426·80	4·12	4·21
2-Me-thiacyclopentane	102	42·40	443·22	9·57	9·76
3-Me-thiacyclopentane	102	87·90	879·15	10·00	10·20
Thiacycloheptane	116	79·90	1163·05	6·87	7·98
2-Me-thiacyclohexane	116	52·40	1154·39	4·54	5·28
3-Me-thiacyclohexane	116	75·80	1049·90	7·22	8·39
4-Me-thiacyclohexane	116	100·00	1308·80	7·64	8·86
2-*cis*-5-DiMe-thiacyclopentane	116	37·80	506·32	7·47	8·68
2-*trans*-5-DiMe-thiacyclo-pentane	116	37·60	492·99	7·63	8·86

References

Ahlquist, L., Ryhage, R., Stenhagen, E. and von Sydow, E. (1959). *Ark. Kem.* **14**, 211.

Antkowiak, W., Apsimon, J. W. and Edwards, O. E. (1962). *J. Org. Chem.* **27**, 1933.

Antonaccio, L. D., Pereira, N. A., Gilbert, B., Vorbrueggen H., Budzikiewicz, H., Wilson, J. M., Durham, L. J. and Djerassi, C. (1962). *J. Am. chem. Soc.*, **84**, 2161.

A.P.I. Research Project 44. Pittsburg.

von Ardenne, M. and Tümmler, R. (1958). *Naturwissenschaften* **45**, 414.

von Ardenne, M., Steinfelder, K. and Tümmler, R. (1961). *Angew. Chem.* **73**, 136.

von Ardenne, M., Steinfelder, K. and Tümmler, R. (1962). *Z. phys. Chem.* **220**, 105, 2nd Ed.

Aston, F. W. (1942). "Mass Spectra and Isotopes", p. 80. Arnold, London.

Barnard, G. P. (1953). "Modern Mass Spectrometry." Institute of Physics, London.

Beckey, H. D. (1963). "Advances in Mass Spectrometry", ed. by R. M. Elliott, Vol. 2, p. 1. Pergamon Press, Oxford.

Bender, M. L. (1951). *J. Am. Chem. Soc.* **73**, 1626.

Bender, M. L. Matsui, H., Thomas, R. J. and Tobey, S. W. (1961). *J. Am. chem. Soc.* **83**, 4193.

Bennett, W. H. (1949). *Instruments* **22**, 472.

Bennett, W. H. (1950a). *Phys. Rev.* **79**, 222.

Bennett, W. H. (1950b). *J. appl. Phys.* **21**, 143.

Bennett, W. H. (1953). *Nat. Bur. Stand. Circ.* **522**, 111.

Berry, C. E. (1949). *J. chem. Phys.* **17**, 1164.

Beynon, J. H. (1960). "Mass Spectrometry and its Application to Organic Chemistry", Elsevier, Amsterdam.

Beynon, J. H. and Williams, A. E. (1963). "Mass and Abundance Tables for Use in Mass Spectrometry." Elsevier, Amsterdam.

Beynon, J. H., Clough, S. and Williams, A. E. (1958). *J. sci. Instrum.* **35**, 164.

Beynon, J. H., Lester, G. R. and Williams A. E. (1959). *J. phys. Chem.* **63**, 1861.

Beynon, J. H., Saunders, R. A. and Williams, A. E. (1963). *Appl. Spect.* **17**, 63.

Biemann, K. (1961). *J. Am. chem. Soc.* **83**, 4801.

Biemann, K. (1962a), "Mass Spectrometry—Organic Chemical Applications", p. 28. McGraw-Hill, New York.

Biemann, K. (1962b). *Angew. Chem.* **74**, 102.

Biemann, K. and McCloskey, J. A. (1962a). *J. Am. chem. Soc.*, **84**, 2005.

Biemann, K. and McCloskey, J. A. (1962b). *J. Am. chem. Soc.* **84**, 3192.

Biemann, K. and Seible, J. (1959). *J. Am. chem. Soc.* **81**, 3149.

Biemann, K. and Spiteller, G. (1962). *J. Am. chem. Soc.* **84**, 4578.

Biemann, K. and Vetter, W. (1960). *Biochem. Biophys. Res. Commun.* **2**, 93.

Biemann, K., Burlingame, A. L. and Stauffacher, D. (1962). *Tetrahedron Letters* 527.

Biemann, K., De Jongh, D. C. and Schnoes, H. K. (1963). *J. Am. chem. Soc.* **85**, 1763.

Biemann, K., Gapp, F. and Seible, J. (1959). *J. Am. chem. Soc.* **81**, 2274.

Biemann, K., Schnoes, H. K. and McCloskey, J. A. (1963). *Chemy Ind.* 448.

Biemann, K., Spiteller-Friedmann, M. and Spiteller, G. (1963). *J. Am. chem. Soc.* **85**, 631.

Breslaw, R. and Gal, P. (1959). *J. Am. chem. Soc.* **81**, 7477.

Brown, R. A. and Gillams, E. (1954). A.S.T.M. E-14 Committee on Mass Spectrometry, New Orleans.

Brubaker, W. M. and Perkins, G. D. (1956). *Rev. sci. Instrum.* **27**, 720.

Bruun, H. H., Ryhage, R. and Stenhagen, E. (1958). *Acta chem. scand.* **12**, 789.

Budzikiewicz, H. and Djerassi, C. (1962). *J. Am. chem. Soc.* **84**, 1430.

Budzikiewicz, H., Djerassi, C. and Williams, D. H. (1964). "Interpretation of Mass Spectra of Organic Compounds", p. 117. Holden-Day, San Francisco.

Budzikiewicz, H., Wilson, J. M. and Djerassi, C. (1962). *Mh. Chem.* **93**, 1033.

Budzikiewicz, H., Wilson, J. M. and Djerassi, C. (1963). *J. Am. chem. Soc.* **85**, 3688.

Caldercourt, V. J. (1958). *J. appl. Spectroscopy* **12**, 167.

Clayton, E. and Reed, R. I. (1963). *Tetrahedron* **19**, 1343.

Collin, J. (1952). *Bull. Soc. roy. Sci. Liège* **21**, 446; **23**, 377.

Collin, J. (1954). *Bull. Soc. chim. Belg.* **63**, 500.

Cope, A. C., Bly, R. K., Burrows, E. P., Ceder, O. J., Ciganek, E., Gillis, B. T., Porter, R. F. and Johnson, H. E. (1962). *J. Am. chem. Soc.* **84**, 2170.

Cottrell, T. L. (1958). "The Strength of Chemical Bonds", 2nd Ed. Butterworths, London.

Dahn, H., Moll, H. and Menassé, R. (1959). *Helv. chim. Acta* **42**, 1225.

De Jongh, D. C. and Biemann, K. (1963). *J. Am. chem. Soc.* **85**, 2289.

Dinh-Nguyen, Ng., Ryhage, R., Ställberg-Stenhagen, S. and Stenhagen, E. (1961). *Ark. Kemi* **18**, 393.

Djerassi, C. (1963). *Pure appl. Chem.* **6**, 575.

Djerassi, C., Budzikiewicz, H. and Wilson, J. M. (1962). *Tetrahedron Letters* 263.

Djerassi, C., Wilson, J. M., Budzikiewicz, H. and Chamberlin, J. W. (1962). *J. Am. chem. Soc.* **84**, 4544.

Djerassi, C., Antonaccio, L. D., Budzikiewicz, H., Wilson, J. M. and Gilbert, B. (1962a). *Tetrahedron Letters* 1001.

Djerassi, C., Brewer, H. W., Budzikiewicz, H., Orazi, O. O. and Corral, R. A. (1962b). *Experientia* **18**, 113.

Djerassi, C., Brewer, H. W., Budzikiewicz, H., Orazi, O. O. and Corral, R. A. (1962c). *J. Am. chem. Soc.* **84**, 3480.

Djerassi, C., Budzikiewicz, H., Owellen, R. J., Wilson, J. M., Kump, W. G., Le Count, D. J., Battersby, A. R. and Schmid, H. (1963). *Helv. chim. Acta* **46**, (3), 742.

Djerassi, C., Budzikiewicz, H., Wilson, J. M., Gosset, J., Le Men, J. and Janot, M-M. (1962d). *Tetrahedron Letters* 235.

Djerassi, C., George, T., Finch, N., Lodish, H. F., Budzikiewicz, H. and Gilbert, B. (1962e). *J. Am. chem. Soc.* **84**, 1499.

Djerassi, C., Budzikiewicz, H., and Wilson, J. M. and Gosset, J., Le Men, J. and Janot, M-M. (1962f). *Tetrahedron Letters* 235.

Doering, W. von E., Taylor, T. I. and Schoenewaldt, E. F. (1948). *J. Am. chem. Soc.* **70**, 455.

Djerassi, C., Nakagawa, Y., Budzikiewicz, H., Wilson, J. M. and Le Men, J., Poisson, J. and Janot, M-M. (1962g). *Tetrahedron Letters* 653.

Djerassi, C., Flores, S. E., Budzikiewicz, H., Wilson, J. M., Durham, L. J., Le Men, J., Janot, M-M., Plat, M., Gorman, M. and Neuss, N. (1962h). *Proc. nat. Acad. Sci. U.S.A.* **48**, 113.

Dugdale, R. C. and Neess, J. C. (1961). Robert A. Taft Sanitary Engineering Center Technical Report W61-3, p. 103.

Eden, J., Burr, B. E. and Pratt, A. W. (1951). *Analyt. Chem.* **23**, 1735.

Elliott, R. M. (1963). *In* "Advances in Mass Spectrometry", Vol. 2, p. 180. Pergamon Press, Oxford.

Falk, G. and Schwering, F. (1957a). *Vakuum-Tech.* **6**, 34.

Falk, G. and Schwering, F. (1957b). *Z. angew. Phys.* **9**, 272.

Field, F. H. and Franklin, J. J. (1957). "Electron Impact Phenomena and the Properties of Gaseous Ions." Academic Press, New York.

Finan, P. A. A., Reed, R. I., Snedden, W. and Wilson, J. M. (1964). *J. chem. Soc.* 5945.

Fox. R. E. and Langer, A. (1950). *J. chem. Phys.* **18**, 460.

Franklin, J. L., Field, F. H. and Lampe, F. W. (1959). *In* "Advances in Mass Spectrometry", ed. by J. D. Waldron, Vol. 1, p. 308. Pergamon Press, Oxford.

Friedel, R. A. and Sharkey, A. G. Jr. (1949). *J. chem. Phys.* **17**, 584.

Friedland, S. S., Lane, G. H. Jr., Longman, R. T., Train, K. E. and O'Neal, M. J. Jr. (1959). *Analyt. Chem.* **31**, 169.

Friedman, L. and Long, F. A. (1953). *J. Am. chem. Soc.* **75**, 2832.

Friedman, L. and Wolf, A. P. (1958). *J. Am. chem. Soc.* **80**, 2424.

Friedmann-Spiteller, M. and Biemann, K. (1961). *J. Am. chem. Soc.* **83**, 4805.

Genge, C. A. (1959). *Analyt. Chem.* **31**, 1750.

Gilbert, B., Brissolese, J. A., Finch, N., Taylor, W. I., Budzikiewicz, H., Wilson, J. M. and Djerassi, C. (1963). *J. Am. chem. Soc.* **85**, 1523.

Gilbert, B., Brissolese, J. A., Wilson, J. M., Budzikiewicz, H., Durham, L. J. and Djerassi, C. (1962a). *Chemy Ind.* 1949.

Gilbert, B., Ferreira, J. M., Owellen, R. J., Swanholm, C. E., Budzikiewicz, H., Durham, L. J. and Djerassi, C. (1962b). *Tetrahedron Letters* 59.

Gillis, R. G. and Occolowitz, J. L. (1963). *J. org. Chem.* **28**, 2924.

Gilpin, J. A. (1959). *Analyt. Chem.* **31**, 935.

Gohlke, R. S. and McLafferty, F. W. (1955). A.S.T.M. E.–14 "Committee on Mass Spectrometry", San Francisco.

Gorman, M., Burlingame, A. L. and Biemann, K. (1963). *Tetrahedron Letters* 39.

Goudsmit, S. A. (1948). *Phys. Rev.* **74**, 622.

Hallgren, B., Ryhage, R. and Stenhagen, E. (1957). *Acta chem. scand.* **11**, 1064.

Hallgren, B., Stenhagen, E. and Ryhage, R. (1958). *Acta chem. scand.* **12**, 1351.

Happ, G. P. and Stewart, D. H. (1952). *J. Am. chem. Soc.* **74**, 4404.

Herzberg, G. (1945). "Infrared and Raman Spectra", p. 193. Van Nostrand, New York.

Hill, H. C. and Reed, R. I. (1963). *Tetrahedron.* **20**, 1359.

Hipple, J. A., Fox, R. E. and Condon, E. U. (1946). *Phys. Rev.* **69**, 347.

Hipple, J. A., Sommer, H. and Thomas, H. A. (1949). *Phys. Rev.* **76**, 1877.

Hipple, J. A., Sommer, H. and Thomas, H. A. (1950). *Phys. Rev.* **78**, 332.

Hofmann, A. W. (1851a). *Justus Liebigs Annln Chem.* **78**, 253.

Hofmann, A. W. (1851b). *Justus Liebigs Annln Chem.* **79**, 11.

Homer, J. B., Lehrle, R. S., Robb, J. C., Takahasi, M. and Thomas, D. W. (1963). *In* "Advances in Mass Spectrometry", ed. by R. M. Elliott, Vol. 2, p. 503. Pergamon Press, Oxford.

Hoover, H. W. and Washburn, H. W. (1940). American Institute of Mining and Metallurgical Engineering Technology Publication No. 1205.

Hurzeler, H., Inghram, M. G. and Morrison, J. D. (1958). *J. chem. Phys.* **28**, 76.

Inghram, M. G. and Gomer, R. (1954). *J. chem. Phys.* **22**, 1279.
Inghram, M. G. and Gomer, R. (1955). *Z. Naturf.* **10a**, 863.
Ireland, R. E. and Newbould, J. (1962). *J. org. Chem.* **27**, 1934.
Ireland, R. E. and Schiess, P. W. (1963). *J. org. Chem.* **28**, 6.
Johnsen, S. E. J. (1947). *Analyt. Chem.* **19**, 305.
Kelly, W., Reed, R. I. and Reid, W. K. (1962). I.U.P.A.C. Symposium on Natural Products, Brussels.
Kendrick, E. (1963). A.S.T.M. E–14 Committee on Mass Spectrometry.
Lampe, F. W., Franklin, J. L. and Field, F. H. (1957). *J. Am. chem. Soc.* **79**, 6129.
Lederberg, J. (1964). "Computation of Molecular Formulas for Mass Spectrometry." Holder-Day, San Francisco.
Lennard-Jones, J. and Hall, G. G. (1952), *Trans. Faraday Soc.* **48**, 581.
Levy, E. J. and Stahl, W. H. (1957). A.S.T.M. E–14 Committee on Mass Spectrometry, New York.
Lund, E., Budzikiewicz, H., Wilson, J. M. and Djerassi, C. (1963a). *J. Am. chem. Soc.* **85**, 941.
Lund, E., Budzikiewicz, H., Wilson, J. M. and Djerassi, C. (1963b). *J. Am. chem. Soc.* **85**, 1528.
Lynch, J. F., Wilson, J. M., Budzikiewicz, H. and Djerassi, C. (1963). *Experientia* **19**, 211.
McDowell, C. A. and Warren, J. W. (1951). *Disc. Faraday Soc.* **10**, 53.
McKinley, J. B. (1957). *In* "Catalysis", ed. by P. H. Emmett, Vol. V, p. 426. Reinhold, New York.
McLafferty, F. W. (1955). A.S.T.M. E–14 Committee on Mass Spectrometry, San Francisco.
McLafferty, F. W. (1956a). *Analyt. Chem.* **28**, 306.
McLafferty, F. W. (1956b). A.S.T.M. E–14 Committee on Mass Spectrometry, Cincinnati.
McLafferty, F. W. (1957). *Analyt. Chem.* **29**, 1782.
McLafferty, F. W. (1963). "Mass Spectrometry of Organic Ions." Academic Press, New York.
McLafferty, F. W. and Gohlke, R. S. (1959). *Analyt. Chem.* **31**, 2076.
Melton, C. E. (1963). *In* "Mass Spectrometry of Organic Ions", ed. by F. W. McLafferty, pp. 35, 163. Academic Press, New York.
Melton, C. E., Ropp, G. A. and Martin, T. W. (1960). *J. phys. Chem.* **64**, 1577.
Meyerson, S. (1959). *Analyt. Chem.* **31**, 174.
Meyerson, S. and Rylander, P. N. (1957). *J. chem. Phys.* **27**, 901.
Meyerson, S. and Rylander, P. N. (1958). *J. phys. Chem.* **62**, 2.
Mitchell, J. J. (1950). "Physical Chemistry of Hydrocarbons", Vol. 1, p. 83. Academic Press, New York.
Mohler, F. L., Bloom, E. G., Lengel, J. H. and Wise, C. E. (1949). *J. Am. chem. Soc.* **71**, 337.
Momigny, J. (1955). *Bull Soc. chim. Belg.* **64**, 144, 166.
Nakagawa, Y., Wilson, J. M., Budzikiewicz, H. and Djerassi, C. (1962). *Chemy Ind.* 1986.
Natalis, P. (1963). *Bull Soc. chim. Belg.* **72**, 264, 374, 416.
Newman, M. S. (1956), "Steric Effects in Organic Chemistry". Wiley, New York.
Otvos, J. W. and Stevenson, D. P. (1955). *J. Am. chem. Soc.* **78**, 546
Olivier, L., Levy, J., Le Men, J., Janot, M-M., Djerassi, C., Budzikiewicz, H., Wilson, J. M. and Durham, L. J. (1963). *Bull Soc. chim. Fr.* 646.
O'Neal, M. J. Jr. and Wier, T. P. (1951). *Analyt. Chem.* **23**, 830.

Orr, D. E. and Wiesner, K. (1959). *Chemy Ind.* 672.
Pelah, Z., Kielczewski, M. A., Wilson, J. M., Ohashi, M., Budzikiewicz, H. and Djerassi, C. (1963). *J. Am. chem. Soc.* **85**, 2470.
Plat, M., Le Men, J., Janot, M-M., Budzikiewicz, H., Wilson, J. M., Durham, L. J. and Djerassi, C. (1962a). *Bull Soc. chim. Fr.* 2237.
Plat, M., Le Men, J., Janot, M-M., Wilson, J. M., Budzikiewicz, H., Durham, L. J., Nakagawa, Y. and Djerassi, C. (1962b). *Tetrahedron Letters* 271.
Plat, M., Manh, D. D., Le Men, J., Janot, M-M., Budzikiewicz, H., Wilson, J. M., Durham, L. J. and Djerassi, C. (1962c). *Bull Soc. chim. Fr.* 1082.
Prosser, T. J. and Eliel, E. L. (1957). *J. Am. chem. Soc.* **79**, 2544.
Reed, R. I. (1958). *J. chem. Soc.* 3432.
Reed, R. I. (1960). *Fuel* **XXXIX**, 341.
Reed, R. I. (1963). *In* "Mass Spectrometry of Organic Ions", ed. by F. W. McLafferty, p. 637. Academic Press, New York.
Reed, R. I. and Reid, W. K. (1963). *Tetrahedron* **19**, 1817.
Reed, R. I. and Shannon, J. S. (1960), I.U.P.A.C. Conference, Sydney.
Reed, R. I., Reid, W. K. and Wilson, J. M. (1962). A.S.T.M. E–14 Committee on Mass Spectrometry.
Rosenstock, H. M. and Melton, C. E. (1957). *J. chem. Phys.* **26**, 314.
Ryhage, R. (1960). *Ark. Kemi* **16**, 19.
Ryhage, R., Ställberg-Stenhagen, S. and Stenhagen, E. (1959a). *Ark. Kemi* **14**, 247.
Ryhage, R., Ställberg-Stenhagen, S. and Stenhagen, E. (1959b). *Ark. Kemi* **14**, 259.
Ryhage, R. and Stenhagen, E. (1960). *J. Lipid Research* **1**, 361.
Ryhage, R. and Stenhagen, E. (1963). *In* "Mass Spectrometry of Organic Ions", ed. by F. W. McLafferty, p. 400. Academic Press, New York.
Rylander, P. N. and Meyerson, S. (1956). *J. Am. chem. Soc.* **78**, 5799.
Sandoval, A., Walls, F. and Shoolery, J. N. and Wilson, J. M., Budzikiewicz, H. and Djerassi, C. (1962). *Tetrahedron Letters* 409.
Schnoes, H. K., Burlingame, A. L. and Biemann, K. (1962). *Tetrahedron Letters* 993.
Shapiro, R. H., Wilson, J. M. and Djerassi, C. (1963). *Steroids* **1**, 1.
Sharkey, A. G., Jr., Friedel, R. A. and Langer, S. H. (1957). *Analyt. Chem.* **29**, 770.
Sharkey, A. G., Jr., Shultz, J. L. and Friedel, R. A. (1956). *Analyt. Chem.* **28**, 934.
Sommer, H., Thomas, H. A. and Hipple, J. A. (1951). *Phys. Rev.* **82**, 697.
Stephens, W. E. (1946). *Bull Am. phys. Soc.* **21**, 22.
Stevenson, D. P. (1951). *Disc. Faraday Soc.* **10**, 35.
Svec, H. (1965). *In* "Mass Spectrometry", ed. by R. I. Reed, p. 233. Academic Press, London.
Thomas, B. W. and Seyfried, W. D. (1949). *Analyt. Chem.* **21**, 1022.
Tickner, A. W. and Lossing, F. P. (1950). *J. chem. Phys.* **18**, 148.
Tickner, A. W. and Lossing F. P. (1951). *J. Phys. Colloid Chem.* **55**, 733.
Whitesitt, J. E. (1961). "Boolean Algebra and Its Applications", p. 25. Addison-Wesley, Reading, Mass.
Wiley, W. C. (1956). *Science* **124**, 817.
Wiley, W. C. and McLaren, I. H. (1955). *Rev. sci. Instrum.* **26**, 1150.
Williams, D. H., Wilson, J. M., Budzikiewicz, H. and Djerassi, C. (1963). *J. Am. chem. Soc.* **85**, 2091.
von Zahn, U. (1963). *Rev. sci. Instrum.* **34**, 1.

Author Index

Numbers in italics are the pages on which the references are listed.

Subject Index

U

V

W

Chemical Formula Index

Chemical Compound Index